Asian Hornet

The Beekeeper's Guide to Defences against the
Yellow-legged Hornet Vespa Velutina

Andrew Durham

Asian Hornet
The Beekeeper's Guide to Defences against the Yellow-legged Hornet
Vespa Velutina

Published 2025 by

Northern Bee Books,

Scout Bottom Farm,

Mytholmroyd,

West Yorkshire

HX7 5JS (UK)

Tel: 01422 882751

Fax: 01422 886157

www.northernbeebooks.co.uk

ISBN 978-1-912271-97-9

Design and artwork DM Design and Print

Cover photograph — reproduced under Extended Licence —

'Real Vespa Velutina Macro' — Eduardo Gonzalez stock adobe.com

Dedicated to:

Les Apiculteurs Française

Contents

Glossary of Abbreviations/Names

AAVO – Friends of the Bees of the Val D'Oise

Abeilles et Fleurs – The magazine of UNAF

ADAAQ – Association for the Development of Apiculture in Aquitaine

ANSES — French Agency for Food, Environmental and Occupational Health & Safety (ANSES)

Apiculteur / Apicultrice —Beekeeper

Apiculture – Beekeeping

BBKA — British Beekeepers Association

Commune — lowest level of French local government, similar to UK parish but with more powers.

CNRS — Centre National de la Recherche Scientifique

EEA — European Environment Agency

ESA — La Plateforme nationale d'Épidémiosurveillance en Santé Animale

FDGDON — Departmental Federation for the Defence against Harmful Organisms (FREDON for regional organisation)

FNOSAD — National Federation of Departmental Beekeeping Health Organizations – Bee Health

GDSA — French Departmental Group for Bee Health

Harpe electrique — a grid of electrified wires in a frame that stuns hornets flying into it

INRAE (formerly INRA) —National Research Institute for Agriculture, Food and Environment

ITSAP — French Institute for the Study Of Bees and Pollinisation

L'Abeille — The magazine of the SNA

Lutte collective — French Department's Nest Destruction Scheme

MNHN — French National Natural History Museum (Museum National de l'Histoire Naturelle)

Municipality — see Commune

Museliere — a cage or muzzle fitted to the front of a hive to keep hornets away

NBU — National Bee Unit

OMF — Open Mesh Floor

SANVE/Tartago — Spanish firm producing electric harps etc.

SNA — National Syndicate of Beekeeping (Syndicat National D'Apiculture)

UNAF — National Union of French Beekeeping (Union Nationale de L'Apiculture Française)

1 - Introduction

The Asian hornet (*Vespa velutina nigrithorax*) or 'Yellow-legged hornet', as it is increasingly called in order to enable the public to distinguish it from other Asian hornets, is a prolific and resilient predator with a remarkable reproductive capacity.

In Belgium, as in France, Spain and Portugal, the Asian hornet has been declared to be beyond eradication. Beekeepers have had to accept that as its numbers continue to grow, the hornet will be in their apiaries predating their honey bee colonies.

In the Flanders region of Belgium (a region that has twice the area of Devon), the reported Asian hornet nest numbers rose from 120 to over 6,400 nests in less than five years, despite a determined campaign of nest destruction. That represented a reported nest density in Flanders of 1 nest in every 2sq.km.; still half the density of nests in some departments in France!

At the end of 2024 the Flemish Beekeeping Institute reported that 93% of the 446 beekeepers responding to their survey, had noticed predation by the hornet on their colonies. Of those beekeepers, 39% reported the hornet entering their hives in the autumn. Why had they allowed that to happen? Why had they not used hornet guards?

Vincent Pellegry, a beekeeper in the Oise Department (Northern France):

As reported in Le Parisien 27 October 2018:

'I never thought I would ever have Asian hornets. The predator arrived in the Oise 4 years ago. We didn't take it seriously enough. Hundreds of hives are impacted by this destructive scourge from Asia. In my apiary, 20 to 25 hives are affected by attacks of Asian hornets. At this rate it's carnage.' says Vincent.

Representing 400 beekeepers, Marc Vallee the President of the Union of Oise Beekeepers says *'Not a single apiary is spared. There are 3 or 4 nests in every commune; probably several thousand nests in the Oise'*

(Oise Department – 686 communes)

Colony losses due to the hornet are higher during the hornet's establishment phase because beekeepers are not prepared for the hornet, don't understand what is happening, or do not know what to do.

The beekeeper can best prepare for the hornet's arrival by understanding the threat to their bees, the interaction between the hornet and the honey bee, and knowing why, when and how to deploy defences. No single defence is a solution in itself, so it is very important to understand where the compromises lie and the correct combination of defences.

My aim in writing this book is to put the necessary knowledge to deal with the hornet into the hands of beekeepers before the initial tranche of colony losses occurs. The book has, of necessity, to look ahead to a time when the hornet is established in the UK.

If this bee escapes the clutches of the hornet, as some do, it will have to learn to fly straight to the safety of the guard bees. If the beekeeper is not to lose colonies to the hornet, they will have to learn to defend their bees. Colonies are very vulnerable in the hornet's establishment phase.
Photograph reproduced under licence – Lothar Lenz stock.adobe.com

We need to learn from the French beekeepers, who have been fighting the hornet for over 20 years and we need to avoid falling into the same pitfalls that even now are hindering them, especially the continuing argument over the spring trapping of foundress queens.

The sudden rise in nest numbers in the UK mainland that was seen in 2023 is a warning of what is to come. The decline in reported nest numbers in 2024 has been attributed to public vigilance and prompt action by our NBU (National Bee Unit) but whilst that undoubtedly helped, the critical factor was more likely the weather.

We'll look at these factors later in the guide but the essential point to grasp is that even if we eradicate the hornets that are here now, more will just keep coming until they do get a foothold. The longer we can put that off the better but ultimately it is a battle we will lose while hornet numbers are so high just across the English Channel.

A traditional beehouse in the Haut-Savoie – A Cabane Grillagee might be the only option (see Section 12.4.6).
Photograph reproduced under licence – Richard Villalon Adobe stock.com

I am acutely aware that what is practical and affordable for the hobby beekeeper in the UK may seem quite impractical for a beekeeper whose beekeeping is on scale that requires palletized hives transported to site with telehandlers to unload them; or whose traditional beekeeping is so

very different from our own. A scale of defences that would cover
4 hives in the hobbyist's apiary may have to cover twice that number in the
professional beekeeper's apiary. I look at this further in Sections 2, 9 & 12.
Where beekeeping may be so marginal that any extra expense is difficult
to contemplate, the fact is that the hornet does not care whether the
beekeeper can afford it or not.

Migratory beekeeping in Provence – more difficult to defend. Photograph reproduced
under licence Yulia Adobe stock.com

I have looked at the debate over the spring trapping of foundress queens
in two quite different contexts. Firstly, in section 5 as part of wide area
campaigns and secondly, in section 10 as part of beekeeper defences in
the area around an apiary. The former involves the general public, whilst
the latter is implemented by the beekeeper. The arguments for and
against each should not be conflated because they are different.

This is a book for the beekeeper. I don't leave the reader with more
debate and questions than answers, which is a feature of some books
on the subject. I use the names Asian hornet, *Vespa velutina,* and Frelon
Asiatique (the French name for the hornet) interchangeably in the text; the
reader needs to be familiar with all these names. I occasionally lapse into

French terminology, just as a reminder that it is the French who have gone before us and it was the French beekeepers who invented and proved the efficacy of the defences covered.

On occasion it is necessary to draw attention to failures in beekeeping practice. Where this is done, it is not done to be critical but only so we learn from our fellow-beekeepers. In the fraternity of beekeeping, the honey bees' needs must come before beekeeper sensitivities.

1.1 - The Beekeeper's Primary Objective.

If the Asian hornet becomes established in the UK, I would suggest that the beekeeper's primary objective should be to keep colonies that over-winter successfully, quickly build up in the spring into strong, productive, healthy colonies, that can survive the Asian hornet's predation in the late summer/autumn, at a cost in time, effort, and resource that the beekeeper can sustain.

Beekeepers whose colonies are too weakened by the hornet's predation to survive winter in good shape or even to survive at all, spend the following spring replacing or building up colonies that do not achieve optimum strength in time for the next attack. Thus, those beekeepers are expending time, money and resource only to see their colonies fall prey to the hornet in the autumn. It is a vicious circle that the beekeeper cannot sustain.

To achieve the objective, the beekeeper will need to:

Be a good beekeeper! You don't need to be a Master beekeeper or have a National Diploma but you do have to be a competent beekeeper. Good beekeeping is a big part of the battle against the hornet.

Be prepared to educate themself - about the hornet, its lifecycle, its interaction with the bee, and defences against the hornet.

Understand that this predator is manageable. A lot of French beekeepers are managing the hornet successfully and thanks to them, we know the answers in advance; they didn't and had to find out the hard way.

Be in the apiary! The Asian hornet is a problem <u>for</u> beekeepers but it is not <u>the</u> beekeepers' problem to solve. No one else is going to do your job as a beekeeper for you.

Throw the calendar away. The development of the problem over the season is very dependent on the weather and the timing varies from year to year.

Be realistic. The aim is not to lose too many colonies. The beekeeper cannot save them all and will not be able to sustain the effort if they try to do so, and their ability to achieve the objective will be compromised.

Be conscious of the potential damage that can be caused to other insects but not to the extent that the beekeeper is frozen into inaction or opts to use traps that are too selective to do the job properly. Sacrificing honey bee colonies of known benefit for an unknown and unquantifiable saving of our 'biodiversity' risks compromising both.

Be Flexible. Beekeepers are going to have to adjust some beekeeping practices.

Be inventive? We are going to copy and adopt the techniques that the French beekeeping associations endorse. There's no point in re-inventing the wheel, so please put the 'Thomas Edison – Junior Inventor's Kit' away; at least until later.

2 - Background & Sources

You will want to know where the information in this guide comes from and in the process of explaining that I can give the background context.

The Asian hornet had arrived near Agen in the Lot-et-Garonne in 2004 but it would be several more years before a chance conversation with a French beekeeper alerted me to the problem of the hornet's predation on honey bee colonies.

I hadn't even heard of the Asian hornet when my family and I were 'en vacances' in Montignac in the Dordogne only three years later. At that time there was only a handful of nests reported in that small town; little did we realise that only two years later the nest numbers there would soar from 6 to 26 nests in just one year. You can get some idea of the impact of that when you look at the map of nests reported in 2009 (see section 3.4).

Once I had heard about the hornet and realised the danger, I started to keep a close eye on the hornet's expansion in France. Tracking its inexorable progress through the MNHN (Museum National de l'Histoire Naturelle) and using nest reports from the departments. I watched as the hornet made its way towards the English Channel. When, in 2014, the reported nest numbers in the Normandy department of Manche got to what I considered to be established, I felt it inevitable that the hornet would soon hitch-hike its way across on one of the cross-Channel ferries and that it was time for me to research the pest in detail.

My trips to France took on a new purpose as I set out on a quest to find the answers to three questions:

1. What exactly does the Asian hornet do?
2. What could I, as a beekeeper, do to help my bees?
3. What was the scale of the problem going to be when the hornet arrived in my Cambridgeshire apiary?

A cunning plan but not a very good one

My initial plan was to use my regular trips to France as an opportunity to visit apiaries, see the problem for myself and to quiz the beekeepers that were always to be found at the local markets. Surely, they would have the answers to the first two of my questions. I soon discovered that the plan was flawed. Too many that I spoke to knew the problem only too well but not what to do about it. I got used to the look of resignation and the famous 'Gallic shrug'. Interesting to note that a survey by FNOSAD in 2023 found that whilst some 50% of GDSA's (department bee health associations) reported *muselieres* of some kind being used, only 15% reported the use of the *harpe electrique*. Over half of the GDSAs called for more information for beekeepers.

There would be no 'Gallic shrug' here. A honey stall on a French Market in 2024. Photograph reproduced under licence – Henry Saint John – Adobe stock.com

We made a point of visiting areas not just where the hornet was a problem but also where the hornet had arrived but not made any progress. I was curious to discover the reason why. For example, in 2017 I visited the Cote D'Or in the east of France where the hornet had arrived in 2009 but by 2013 had almost disappeared. It was explained to me that the weather wasn't suitable for it and the hail storms of 2012 and 2013, that wrought havoc in the vineyards, had been the final straw.

The same year we visited the market in the town of Dole (Jura) in the far east of France and spoke to beekeepers but again it was a case of the

Gallic shrug. Ironically, given the timing of our visit, that was the year the hornet officially arrived in the department so I hope they were quick learners. They certainly needed to be because the surge of 2022 caught the east of France off guard and many beekeepers lost colonies.

In apiaries, I saw that defences that I already knew about were either not being deployed or had been used inappropriately. It was also evident that defences that had been developed years before in the departments where the hornet had first arrived, had not found their way into departments further north. It puzzled me to see newspaper articles featuring this or that beekeeper holding a new trap that they had 'invented' to deal with the hornet, when I knew the same existed further south. As the hornet spread north there was an element of each department reinventing the wheel. It was the lack of knowledge on the part of some beekeepers and constant reinvention that set me off on the BBKA talks circuit and writing articles for the BBKA News, ultimately culminating in this book. I didn't want the same problem occurring in the UK.

Whilst it was great fun spotting the tell-tale sign that an apiary was hidden in a wood off the beaten track and racing off to study what was happening in it, I determined that individual beekeeper experiences and observations (including my own) although useful, could be inconsistent and I was never sure how representative they were.

Photograph by the author – 'racing off' to find beehives hidden in a wood near Dijon.

Furthermore, there was a lot of inaccurate exaggeration and misinformation surrounding the hornet, especially on social media, so understanding the scale of the problem and its impact on beekeeping was just as important as finding out which defences worked best. I wasn't going to get that information the way I was going about it. I needed to identify where the consensus lay and that meant I needed to get a much wider viewpoint.

A wider viewpoint

I soon found that I got a good view of what was happening in a department from the departmental beekeeping associations, the equivalent of our county associations. I looked at the advice that they were giving members on how to deal with the hornet. Newsletters and chairperson's reports proved very informative. Claims that the hornet had caused severe losses in one department were put into perspective when they were also attributed by the association's chairperson to high winds and a drought.

Press reports were generally not to be relied on. Nevertheless, a story about heavy winter colony losses in the Haut-Vienne over the winter of 2017/18 could not be ignored. The interview was with a beekeeper who lost 28 hives, Helene Delaplace from Eymoutiers, and some of her fellow beekeepers. They said the losses were due to a lack of forage, pesticides, varroa and the fact that beekeeping had become so much more difficult. I took an absence of any mention of the hornet as an indication that even though it was definitely present in the department, it obviously wasn't a big problem for beekeepers; although I think the story would have been different a year later, as there was a hornet surge in 2018.

An invaluable source of information was the reports from the FDGDON / FREDON (Departmental Federation for the Defence against Harmful Organisms (FREDON for regional organisation)), a national state-accredited pest and disease organisation operating in the agricultural sector. As the scheme managers for many Departmental Nest Destruction Schemes, they coordinated the nests reports from local town halls (the 'Marie') with those from the local 'referents' (often beekeepers) who checked the reports and followed up with the pest controllers who destroyed the nests.

The FDGDON reports were fastidious and very detailed . In addition to their commentary on the situation, the timing of nests reports etc., they usually contained maps showing the number of nests reported in each municipality which I was able to compare with population density maps.

The reported nests numbers from the departments were correlated with weather reports from Meteo France and more recently, for Belgium , the Belgian Met Office. Variations in nest density were correlated with climate and environmental mapping from the EEA's (European Environment Agency) databases.

Lutte collective (French Department's Nest Destruction Scheme) publicity poster courtesy of FDGDON 35 Ille-et-Villaine. (Photograph Eduardo Gonzalez stock adobe.com).

To get a national overview, I subscribed to three national beekeeping organisations in France:

▶ UNAF — Union Nationale de L'Apiculture Française, the French beekeeping union. A union that has struggled with a lack of government support from the outset of the invasion and that has been vociferous in its condemnation of the interference from well-meaning ecologists some of whom now realise that they should have acted earlier. Nevertheless, I was surprised that it took so many years for UNAF to get its own Special Edition on defence out

> to beekeepers (no. 5 – see Further Reading)) A more recent guide is available on-line (no. 6 — see Further Reading).

- ▸ SNA – Syndicat National D'Apiculture, the national beekeeping syndicate. In addition to articles on the Asian hornet, membership of the SNA was useful to learn what had been happening from their regional reports and it helped to put beekeeping into perspective. The monthly 'Edito', from the president of the SNA, opened my eyes to a multitude of problems faced by the French beekeepers across the English Channel. Although we often face the same ones here in the UK, it is clear that some of those problems are more severe in France.

- ▸ FNOSAD — the national organisation of bee health groups. They keep members up to date with the latest news on the Asian hornet and report on surveys of their member associations on subjects such as, how many beekeepers are using which defences against the hornet and to what effect.

Whereas in the UK the majority of beekeepers are hobbyists, in France the majority do it either as their main source of income or a much-needed source of supplementary income. As such, beekeepers are under enormous pressure from many different directions, adulterated or cheap imports, increasingly intensive agriculture using more and more pesticides to name but a few. Losses due to the hornet and the cost of defences just added to the list.

 The need for beekeeper education cannot be over-stated. As chance would have it and just as I was finalising this book, I noticed a question posed by a beekeeper in the Q&A pages of the December 2024 issue of *L'Abeille* (the National Syndicate of Beekeeping (SNA) magazine). Jean-Luc posted a photograph of over 35 Asian hornets attacking the entrance to one of his hives and asks the professional beekeeper 'Que faire?' – What to do?

The advice given to Jean-Luc was to move his hives, that predation had been very late in some areas and measures such as the use of *'muselieres'* (muzzles) hadn't been enough. Clearly Jean-Luc hadn't realised, and the professional beekeeper answering the question hadn't noticed, that he was using a mouse-guard to protect his bees when he should have used a hornet-guard. It was no barrier to the hornet at all. You will read more about the difference between the two in the Section on Defences in the Apiary but this was just one more illustration of a very common error that results in beekeepers losing colonies in the autumn.

Something that the French have that we do not, are the separate GDSA's (Departmental Bee Health Associations) who work under the aegis of FNOSAD and GDS France. They are much more than the equivalent of our bee inspectors and another very useful source of information. Some GDSA's act as scheme managers in nest destruction campaigns and even carry out nest destruction.

Media reports of mass attacks by the hornet on runners, walkers and cyclists were followed up with the local SDIS, the organisation responsible for fire and rescue. This included the pompiers who were often called to deal with such incidents and who bore the brunt of the initial response to rapidly increasing reports of nests from a worried public. Their professionalism and dedication to the task was truly admirable. One such report from a local firefighter commander, who led a response to a mass attack by 'Asian hornets', quickly confirmed that the culprits were, on that occasion, European hornets.

Nevertheless, one incident in Deux-Sevres that involved a mass attack on primary school children saw 10 ambulances dispatched along with 41 first responders and a specialist team to destroy the nest. Although the hornet had become a fact of daily life in many areas, such incidents as this drove home the impact of the hornet on society.

The ESA's (La Plateforme nationale d'Épidémiosurveillance en Santé Animale) annual national survey of honey bee colony losses provided invaluable information on winter loss rates from 2018 onwards, which I tied in with reports from individual departmental nest destruction campaigns and reports from individual department beekeeping associations. Not only did it put annual winter losses into perspective, it confirmed INRA's (National Research Institute for Agriculture) early insistence that honey bee colony losses were too multifactorial to separate out those attributable to the hornet.

Furthermore, winter losses in some departments that were heavily infested were seen to be consistently lower than the national average, suggesting that the beekeepers in those departments were successfully managing the threat from the Asian hornet. Whereas loss rates increased in the sudden surge of 2022 in departments that had previously had low nest densities.

Finally, there are the research papers on the hornet, there are now over 200, but I restrict myself to those directly relevant to beekeeping. I have listed almost 70 for further reading, and where I have referred to a particular paper in the text I have put the corresponding number where it will be detailed in 'Further Reading'.

I haven't followed the normal referencing protocols where the scientist is listed before the subject, presumably because scientists want to know what a particular scientist has published. I am more interested in the subject of the paper and have put that first. I have also tried to tie the papers to the sections, where that is possible. I think that is more meaningful than a list of scientists that many will never have heard of.

This book represents the collective knowledge of a whole range of beekeepers, associations and researchers across France and Spain. It is a collation and a distillation of authoritative sources of information. On balance and whilst not underestimating the value of first-hand accounts, I think that is more reliable than any single beekeeper's experience regardless of how many hornets they have had in their particular apiary.

Destruction is not a simple action and the utmost caution is necessary.
Only authorized and properly equipped people should act on the ground.
If this hornet is, when isolated, very peaceful, the voluntary or involuntary approach of the nest can induce a collective attack that presents a real danger to the exposed person.

Jacques Blot – Directeur Technique ADAAQ

(Association for the Développement of Apiculture in Aquitaine)

3 - Thinking Well Ahead of the Hornet's Arrival

3.0 - Introduction

As of 2024 the hornet's incursion is mainly limited to the far south of England but you should think ahead in case you are setting up a new apiary or perhaps thinking of re-organising an existing one. As you can't just shuffle hives around like a deck of cards it might pay to plan ahead.

The first two of my questions (what did the hornet do and what could I do about it?) were answered fairly quickly; after all, by that time the beekeepers in the south-west of France had struggled with the hornet for more than 10 years. The third was proving much more difficult. I needed to gain some idea of the scale of the problem or would not be able to plan or resource defences properly. So, in this section we look at some practical aspects and the process I went through to answer that critical third question.

3.1 - Apiary Location

Apiaries in some locations are likely to come under heavier predation than others. Even if you can't choose where your bees are kept, you might want to know the risk and plan accordingly (see below in 3.4).

3.2 - Apiary Layout

Some hive positioning is more vulnerable than others and some layouts can be much more efficient if you have to deploy more advanced defences (refer to section 12.2 on Defences in the Apiary – Environment).

3.3 - Type of Bee

There is evidence to suggest that more defensive types of bee fare better under hornet predation (*Apis mellifera mellifera* for example).

Some imported bees/queens are to be avoided:

Apis mellifera ligustica (Italian bee) — are known to overwinter with large colonies, which means that large winter stores are needed and they have a habit of making any additional food into brood. Not good traits to have when the Asian hornet is around. See Further Reading nos.55 & 70 for more on Apis Ligustica's vulnerability.

Apis mellifera caucasia — known to be a gentle bee that doesn't build up quickly in the spring. Not a good choice if we want colonies to build up quickly. So that's two traits we don't need.

Apis mellifera carnica — often described as quiet and gentle but they can get feisty (from experience of my own Carniolans). Reacts to food income changes very quickly. Overwinters well on low stores and builds up quickly in spring but then swarms. Good traits to have in this context, but as a commercial bee, Brother Adam thought it was not a good choice. Perhaps better to avoid that one as well.

An Asian hornet that got too close to the guard bees. *Apis mellifera* just needs time to learn. Note the green Asian hornet restrictor used to prevent entry in the autumn. Photograph reproduced under licence - Shocky stock.adobe.com

Ignore the constant chatter about *Apis cerana*'s ability to cope with the Asian hornet (Mexican waves and balling etc.) and *Apis mellifera*'s inability to defend itself. Apis *mellifera* just needs time to learn. *Apis cerana* is so very different as a bee that it's not worth considering further except perhaps for one thing, and that is its use of dung to put off the Asian hornet's bigger relative, *V. soror*. Although Prof. Heather Mattila, who reported this, didn't observe the same behaviour by *A. cerana* with *V. velutina*, it's worth remembering when thinking about olfactory attractants or repellents.

3.4 - What is the Scale of the Problem Likely to be in My Apiary

The greater the nest density, the more likely it is that your apiary will be affected and the higher the number of nests the worse the problem will be because multiple nests could be predating your bees.

I wanted to get a fix on what influenced nest densities and how evenly those densities might be spread. The departmental nest destruction reports helped to some extent but the difference between urban and rural nest densities was huge. If the French scientists were right and the hornet is an urban-dweller then I wouldn't have much of a problem in my rural apiaries. However, if that were the case why were so many rural beekeepers reporting colony losses? My instinct told me that in the countryside the nests just weren't being reported.

I started by trying to understand why some regions of France were more affected than others.

Why the difference between the south-west and the north west of France?

Studying the reports from the beekeeping associations and from the departmental nest destruction schemes, it was obvious that the problem was worst in south-west and western France. The north-west of France and most of eastern/north-eastern France had far less of a problem.

The French MNHN (Museum National de l'Histoire Naturelle) is the French Government's chief advisor on the Asian hornet. Its website https://frelonasiatique.mnhn.fr/biologie/ shows the output from a computer model that attempts to predict the favourability of areas to the Asian hornet. The model was in many respects just another modelling attempt by scientists to predict the spread of the hornet, but what was immediately apparent was that this particular model was very good at showing the actual density of nests in France, as confirmed by my scrutiny of the nest reports from the French departments.

Why were there so many more nests in some departments than in others? Furthermore, although the hornet had spread to the Hauts-de-France departments of the Somme, Pas-de-Calais, Oise and Aisne by 2016, the nest densities there were very low and the hornet's expansion had almost stalled. Was that the extent of the hornet's expansion? If that were the case, we in the UK would have little to worry about.

It was important to find the answer because although the hornet could hitch-hike from ports in Brittany and Normandy to the UK, the odds of survival for the few hitchhikers that made it across were far less than if the hornet could fly across from north-western France. Furthermore, the cross-Channel ferries from Calais or Dunkerque would be more of a problem because it is a much busier route. It is a simple question of numbers — the more hornets that arrive, the greater the chance that some will succeed in establishing a foothold.

I focused my attention on Normandy because its westernmost department of Manche (with the port of Cherbourg) had notably more nests each year than the other Normandy departments to its east. This was particularly evident up to and including 2019. The D-Day department of Calvados, adjacent to Manche, had fewer nests. You can see this for yourself if you look at the table in section 5.2 on nest destruction. Both Manche and Calvados had well-organised nest destruction schemes and the reports from the FDGDON scheme managers were very detailed. It wasn't obvious why there were so many more nests in Manche when the hornet had arrived in the two departments at about the same time.

A Eureka Moment!

The answer came for me when I discovered a map in the European Environment Agency's library. It showed the observed climatic zones for Europe and the two zones that caught my attention were the Maritime South and the Maritime North.

Prior to 1995 the observed border between the Atlantic Maritime South and North Zones lay well to the south-west but by 2016 it had shifted to include the whole of Brittany, Western Normandy, and it even extended up towards the Ille-de-France. Overlaying the MNHN's model showing what I knew to be actual nest densities, I could see that the Maritime South zone and the MNHN nest density map in the western half of France were a perfect match. It was a 'Eureka' moment. It also explained why the

Channel Islands had so many hornets (they are in the Maritime South).

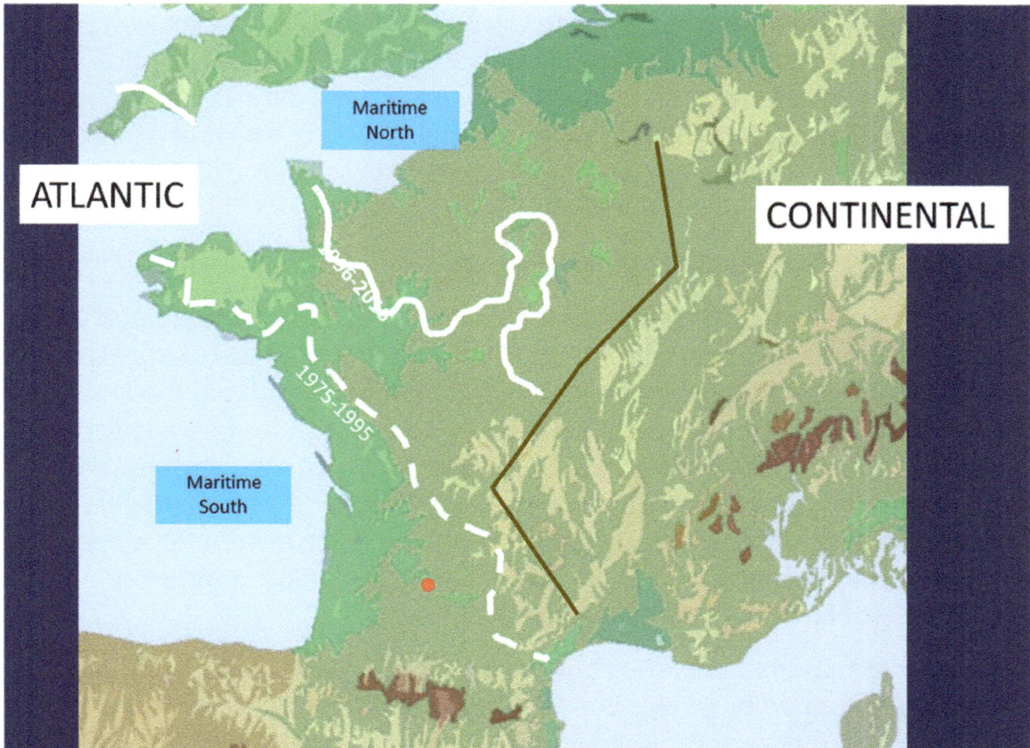

Graphic showing east/west divide and position of maritime zone boundary to 1995 & 2016. The red dot indicates the approximate position of the first nest in 2004.

It is in the Atlantic-influenced Maritime South Zone that we see the highest nest densities. As we progress into the Maritime North Zone, we see a marked reduction.

Why the obvious difference between the east and the west of France?

Across an east / west divide in France caused by the Massif Central and the hills of the Morvan; in eastern France the hornet is much less of a problem (more a nuisance) than in the west. Furthermore, it gets less of a problem the further one moves north.

For example; the eastern region of the Auvergne-Rhone-Alpes, which comprises some 12 departments, was reporting under 3,000 nests annually right up until 2020. Although post the 2022 surge, just one of those twelve departments (Isere) had reported 2382 nests. Even so, that was a third of the nests reported in Manche.

The reason for this disparity lies in the fact that France's climate is largely determined by two major weather systems, the Atlantic (milder-wetter) dominates the west and the Continental (longer colder-winters and shorter-hotter summers) dominates the east, except where the Mediterranean climate intervenes in the south. This difference in climate determines the weather, which plays a major role in the hornet's nest numbers each year (see section 7 - The Asian Hornet – A Consummate Predator – for the reasons for this).

Accelerating Climate Change intervenes

If the situation in northern France and Belgium seemed to have stalled, it wasn't to remain that way for long. In the less than 10 years since the modelling it became evident that both the MNHN model and the observed climate zone maps are being overtaken by an accelerating climate change. The clue to that was a marked surge in nest numbers in north-eastern France and Belgium caused by two years of weather that proved very favourable to the hornet. The climate there had been getting progressively more favourable to the hornet for more than 30 years; the weather in 2021 and 2022 was the kick-start it was waiting for.

In Belgium in 2024, the hornet has continued to gain ground, although its progress was delayed by poor weather conditions in the spring that affected all Hymenoptera. As might be expected, from what I have already described in France, Belgium has more nests in the north and west than in the south and east, which has the hills of the Ardennes and is also in the continental weather zone of influence. France also saw a delayed start in 2024 but with some areas subject to heavy predation later in the year.

In the longer term, the fact is that national meteorological offices in the UK, France and Belgium are all reporting an underlying climatic change towards warmer wetter winters and dryer summers with more intense weather events. Conditions that suit the hornet well; and that Maritime South Zone is still expanding. But for now, we are in the Maritime North Zone.

Urban or Rural?

At French departmental level there are many more nests reported in urban/suburban areas than in rural areas. Reported nest densities in 2020 in the town of Vire (Calvados, Normandy), sitting just inside the Maritime North Zone, were 10 nests per sq.km., not unusual for middle/eastern Normandy. Along one road alone, the Rue des Landes, there were 5 nests in just 250 metres! Away from the urban/suburban areas the reported nest densities were below 1 nest per sq.km., in fact most rural areas were below 0.6 or even zero. It would be wishful thinking to believe there were so few or no nests at all in those areas.

In the adjacent department of Manche, western Normandy and in the Maritime South Zone, nest densities of 14-16 nests per sq. km. in urban/suburban areas were not unusual. This is in a department that has a well-organised nest destruction scheme. Various hypotheses have been offered to explain the urban/suburban densities; island heat effect, availability of food, more suitable habitat for embryo nests under cover, the hornet favouring areas disturbed by human activity etc..

Whilst small differences in temperature have a major effect on Hymenoptera, the differential between urban and rural seemed outweighed by regional temperature differences and if food was so plentiful in urban/suburban areas why was the hornet's diet 70% honey bees in urban areas compared to 30% in rural areas? That doesn't suggest more food in urban areas, although markets have a real problem with hornets on food stalls.

A comparison of '*Lutte collective*' nest report mapping with departmental population maps, an exercise I repeated in a number of departments, clearly shows that reports of nests are directly linked to population density. This phenomenon was also apparent in the early years of the Wallonia nest mapping, where even reported nests in the countryside were so obviously located along roads and next to houses.

Yet in the small Channel Island of Jersey, where the island population is more cohesive and is on high alert, the nest distribution is more evenly distributed and reported nest numbers are much higher (per sq. km. overall) than in the French department of Manche just 20 miles to the east.

In fact, Jersey's 335 reported nests (2023) in 120 sqm km. would represent 16,750 nests in Manche (circa 6,000 sq.km.). But in Manche they reported just 7,700 nests (2023), suggesting that there are many more unreported nests in France than is realised. Furthermore, Spanish researchers have called into question the French insistence that the hornet is an urban dweller, believing that there are just as many nests in the countryside.

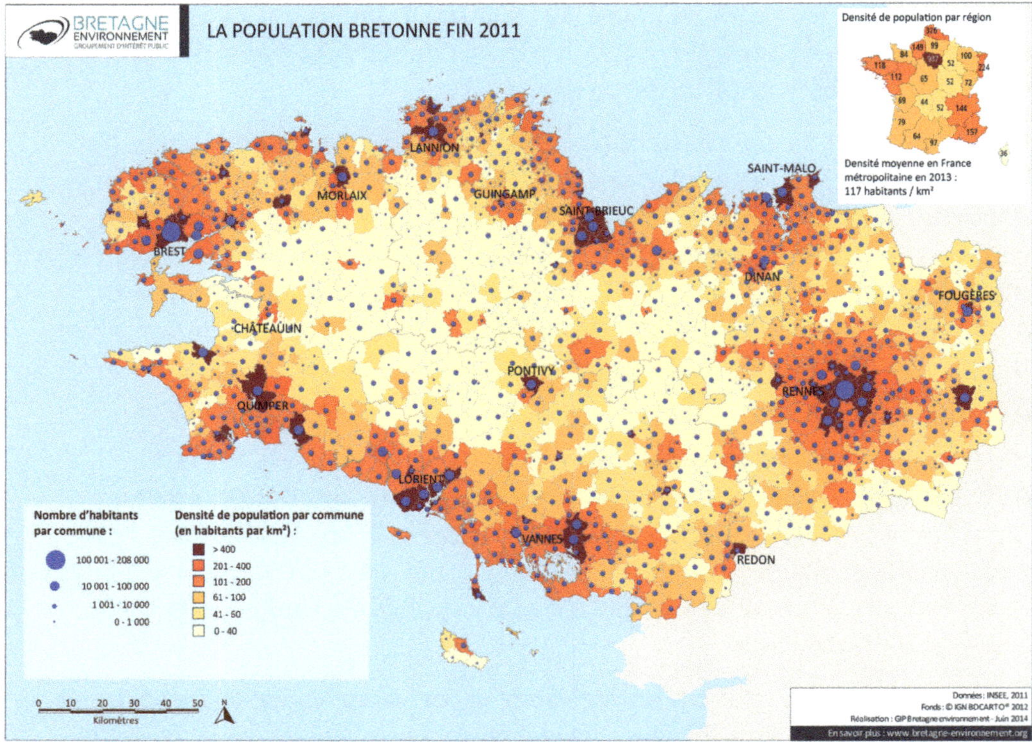

Population density map of Brittany showing the sparsely-populated interior. Comparison of nest report maps with population density maps revealed that the areas with a very low population density always coincided with the areas where very few nests were reported.

Looking at the enormous numbers of foundress queens caught in spring trapping campaigns in Morbihan (Section 10), I concluded that the French have absolutely no idea how many nests they have in the countryside.

I believe the contention that the hornet is an 'urban-dweller' is grossly over-stated and the underlying problem is that urban-dwellers are more likely to detect nests and will report them, whereas the rural nests are more likely to remain hidden and even if they are detected, it is more likely that they will not be reported. Trapping reports from INRA Bordeaux may have been reported to be higher on its urban/suburban site than in a more rural apiary, but there is probably more prey in the countryside and fewer hornets had found the rural apiary.

So far, I had established the difference in nest densities between climatic zones and that there may be more nests in urban areas, even if not to the extent believed, but I live in the countryside. Now I wanted to zoom in on the situation on a more local scale.

How close does a nest have to be in order to be a problem?

Consider the plight of the beekeepers in Montignac (Dordogne) where, in 2009, there were 26 nests in just one small town of c.3000 people.

Asian hornets can predate more than a kilometre from the nest but scientists have set 1km. as the standard radius for the destruction of nests and for spring trapping around apiaries. The radius shown by a dashed blue line in the graphic below is set to 600m, which gives us beekeepers a more realistic radius to work on and because that covers the vast majority of the nest's predation.

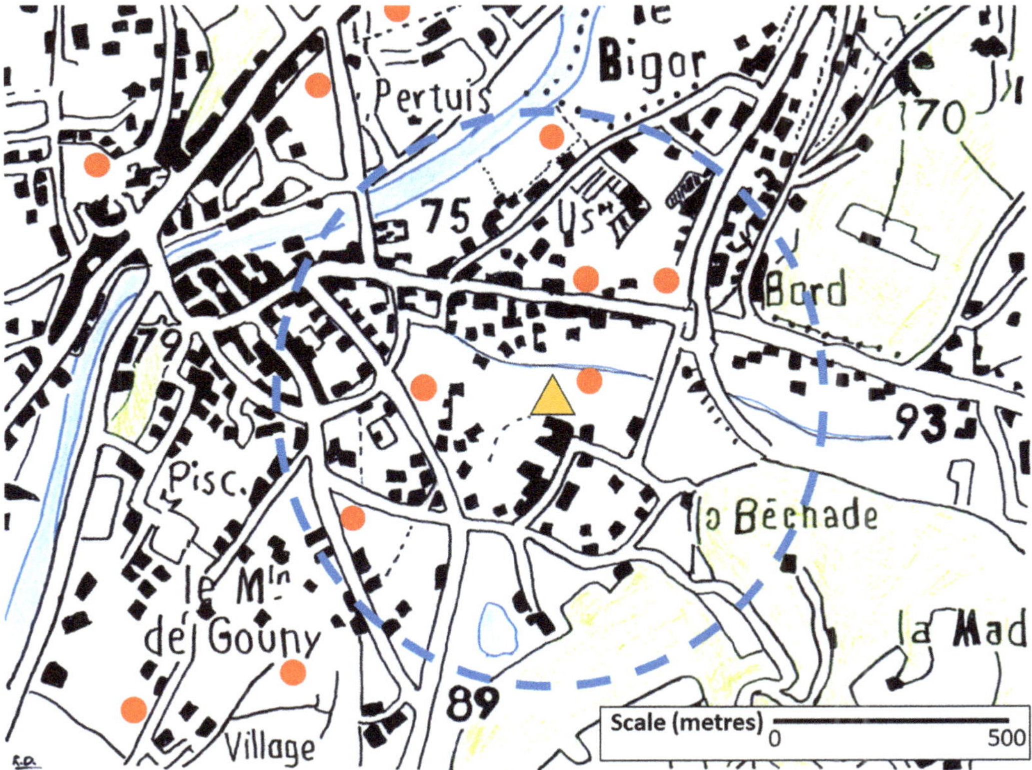

Graphic showing actual nest locations in Montignac (Dordogne) in 2009 – example apiary is shown by a yellow triangle.

3.4.1 - The Situation in the United Kingdom

Given that, for now, the Maritime North Climate Zone covers pretty much all of the UK mainland, will the distribution of the hornet be the same across the whole country? In part it depends on whether the recent acceleration in climate change continues but the MNHN model was certainly predicting that by the year 2100, 95% of the UK would have a 90>% favourability for the hornet. However, climatic zone mapping seems a rough tool for this purpose and a greater subtlety lies in the Environmental Zones mapping. Using those maps and comparing them with the presence of the hornet in France: if the hornet gets established in the south of England, the area of concern would extend up to a nominal line drawn between the Severn and Humber estuaries. Interesting to note that my assessment coincides with that of CARI, the Belgian scientific beekeeping group.

North and west of that line, the higher ground and habitat of Wales and anywhere from the Peak District north, will be much less suitable for the hornet. There may of course be areas that are exceptions to that, such as the Vale of Pickering and Vale of York in the east, the area between the Clwydian hills and the Pennines and, in Wales, the Gower peninsular.

South and east of the line we are likely to see a similar situation to that of Flanders, Belgium, which in 2023 was comparable in nest density terms to Calvados and eastern Normandy (see section 5.2 - Nest Destruction).

The problem will be worse in the far south and south-west of England, especially if the Maritime South Zone continues its expansion (note that the Maritime South Zone border was in Cornwall in 2016).

Although we will see wide variations in nest numbers from year to year due to the weather, I could now be reasonably certain that we would not experience the nest densities suffered in Brittany or western Normandy, at least not for some years. The simpler and cheaper measures (muzzles and entrance restrictors) would therefore most likely suffice, supplemented by large capacity traps if required. The more expensive electric harps would be an option, if necessary. However, even in areas of France that are considered to be infested with the hornet there are wide variations within departments and one apiary can experience very different predation to another just a few miles away.

Some apiaries are going to be more affected due to their location; those that are:

- ▸ Close to rivers and bodies of water – the hornet needs water
- ▸ Urban/suburban – the hornet appears to favour such areas but less so than is often claimed
- ▸ In pastoral/fruit agriculture areas – there is more insect prey to support more/bigger nests

Some apiaries are going to be less vulnerable, those that are:

- ▸ In forests – a much lower percentage of nests is reported in forests, at least in coniferous forests. There seem to be plenty of insects in deciduous forests in France (not a good place to walk in shorts if our experiences are anything to go by!).
- ▸ In intensive arable agriculture areas – fewer insect prey because of less habitat, more pesticides etc.

There are far fewer nests reported in forests. Photograph reproduced under licence - Véro des Cairns Adobe Stock.com

The direction of the prevailing wind during the autumn is important. If the area downwind of the apiary is intensively farmed there will be fewer nests in it than if the apiary is upwind of an area that is very favourable in habitat terms.

You might also consider how many other apiaries there are in your vicinity and where they are in relation to you in terms of the prevailing summer/ early autumn winds. Careless beekeepers in your vicinity will attract the hornet to their apiaries, thus acting as decoys for yours; as long as you don't make the same mistake and draw the hornet's attention to your apiary.

Put together with my assessment of likely nest densities and what that meant with regard to defences; that was as close as I was going to get to 'what would the scale of the problem be in my apiary?'.

How quickly is the threat going to materialise?

The threat will take time to build, perhaps even years. A real problem is that a slow or uncertain build-up, as seen in northern France and more recently, Belgium, can lead to complacency on the part of beekeepers.

The biggest danger comes during a 'surge year' when conditions conspire to cause a sudden and huge increase in hornet nest numbers: as happened in the Oise Department in 2018 and later across most of France and Belgium in 2022.

A bucolic scene in an orchard in Provence but these hives are very vulnerable.
Photograph reproduced under licence – dvoevnore Adobe stock.com.

4 - Science and Solutions

4.0 - Introduction

In the April 2023 edition of 'Abeilles & Fleurs' (UNAF's magazine), an editorial by Christian Pons (President of the French Beekeeping Union) accused scientists of wanting to ride on the wave of interest following the 2022 surge in hornet nest numbers *'in order perhaps to access a European windfall of several million euros'* granted by an EU that was concerned at the sudden increase in the hornet's expansion. What, he demanded, had these *'tout ce beau monde'* (beautiful or fine people) been doing for the past almost 20 years, except try *'by all ways to destabilize and discredit the proposals that only UNAF has developed to alert and contain this predator'*.

A January 2025 report in L'Abeille (SNA's magazine) on the announcement in the USA that *V. mandarinia* had been eradicated, stated:

'The responsiveness of the United States to implement an effective strategy (against Vespa mandarinia) is in total opposition to that which for twenty years was imposed in France by a small cohort of incompetent scientists and technocrats who have kept France in an irresponsible and dangerous immobility in the face of the invasion of the yellow-legged hornet (Vespa velutina)'.

What had led to these extraordinary outbursts of anger and frustration? I have already referred to a rift between scientists and beekeepers and to an ecological correctness that hampered beekeepers for many years in France. The frustration had been building up for years, evident in editorials and reports in the national associations' magazines and resulting in a withdrawal of all cooperation at one point. Too late for some scientists to now question whether the response in France could have been more robust, that water went under the bridge long ago.

Science is everywhere in the fight against the Asian hornet and it comes up again and again in this guide to defences. Some of it is good science but some of it has been based on very incomplete data, using rather questionable methodology and some dubious field research.

Why do we beekeepers need to look at this? Because in a war against the Asian hornet, it is important to know who your allies are and what solutions they might bring to your aid. Many beekeepers look to the scientists both as allies and for advice/solutions; and they delay making preparations to defend their bees because they think something will turn up to solve the problem.

This section looks at what happened in France, why not all scientists were necessarily allies and why those beekeepers waiting for solutions were in for a long wait. I'll leave the fine detail in the relevant sections and concentrate here on the background, with enough detail to give you the general theme. Could it happen here? I hope not but there are definitely lessons to be learnt from the French experience.

It all started in France with the failure of the French Government to do the very thing it was there to do; govern.

4.1 – A Failure of Government

The French government's position was that it could not commit to a policy to deal with the hornet until someone came up with a quantifiable strategy that would work. Nor could it afford to declare the Asian hornet a category 1 health hazard, as it then becomes the government's responsibility to deal with it at the State's expense. It is the same situation even today. Perfectly understandable but it is an abdication of responsibility. At one point a minister actually said that the hornet was a problem for the beekeepers.

Following the hornet surge year of 2022 there were many calls for action and a bill was passed in the French Senate that called for more action in the fight against the hornet. In March 2025 it was formally accepted for legislation in the National Assembly (the other half of the French legislature). The passing of the bill in the Senate received a lot of publicity but what was not so well-publicised was that its contents were heavily diluted in the committee stage and what actually passed in the Senate was not much more than a statement of good intent. It is unlikely that attempts to restore the amended bill to its original state will be successful because it would call for state funding that they cannot afford.

In the years following the hornet's invasion and in the vacuum caused by the lack of a strategy, departments were left to their own devices, struggling to deal with an invasive pest that was affecting ordinary people's lives and the citizens were taking matters into their own hands.

Ecologists in France became alarmed at the public's reaction to the arrival of the Asian hornet where there was a near hysteria (described as such by the State Prefect in one department) that even saw photographs in the press of children in primary schools making bottle traps in which to catch the invasive pest. People were featured proudly holding their 'Asian hornet' traps or in their gardens with hedges adorned with traps, and even the members of a hunter's lodge were pictured with crates of empty plastic lemonade bottles ready to be cut up for traps, eager to get going on a new source of prey. Villages were adorned with posters advertising public meetings to discuss the pest; often addressed by anxious 'apiculteurs'.

Non-selective drowning traps are not suitable for wide-area trapping but can be surprisingly effective – Photograph reproduced under licence – PLG stock.adobe.com

The traps that the public were using were simple drowning traps, cheap, easy to make but surprisingly effective at catching insects. The problem was that the traps were non-selective and catching far more 'other-species' than *Vespa velutina*. Ecologists, often in influential positions, naturally wanted to deal with the threat to biodiversity caused by the public's out-of-control non-selective trapping.

The MNHN, the government's chief advisor on the hornet, told the public not to trap and accordingly that message went out in all departments. The hornet would, it said, quickly reach saturation point when everything would get back in balance! Little did the scientists making that statement know that saturation point would mean an intolerable density of nests; as pointed out by the department of Ille-et-Vilaine almost 20 years after the arrival of the hornet (see section 5.3 - Nest Destruction).

It's not an argument you hear put forward anymore. In fact, the same scientists are now reinventing themselves 'post-2022' as leaders in the new fight against the hornet. But in the years that followed the hornet's invasion, what were the scientists doing? The problem was and still is, that science costs money.

4.2 - Research Funding

Without a government policy and State funds, the only research that could take place was at the initiative of those bodies that could divert funds from within existing and already tight budgets.

Scientific research is a precarious living until the scientist is well-established. Those principal investigators with research teams to support, have to target their applications carefully and they are often submitting proposals to funding programmes that are years in the future.

Applications therefore had to be carefully targeted to the funder's agenda if they were to be successful and little research was undertaken to find solutions for the beekeepers to use in their apiaries. What research did take place took years to get started, years to undertake and even more years to publish.

Other scientists in related fields saw a new target for research income. Researchers started publishing papers, many of which merely summarised the very few research papers that existed on *V. velutina*, and then proposed new avenues for them to research. Avenues that promised biodiversity-friendly solutions. Some were very successful in getting substantial funding. They have been less successful in delivering on their promises.

In France everything was taking too long. It took until November 2018, 14 years after the hornet arrived in France, for the French scientists to confirm that 'the hornet affects foraging and survival probability in honey bee colonies' (no. 48 in Further Reading, and section 8.3 - 'The Hornet & The Honey bee'); the beekeepers' claims about colony losses having been dismissed by scientists as 'un-scientific and anecdotal'. That research at least gave scientific credibility to the beekeepers' claims but even then it lacked sufficient scope, was under-funded and under-resourced.

In sharp contrast, the Spanish response has been much quicker with local authorities continuing to support nest destruction, even introducing novel methods of nest destruction (in Galicia and Asturias nests may literally be blown up with gunpowder squibs). Researchers have been quick to get involved, papers have not only been published much more quickly but also target practical matters. It is notable that the first French paper on the '*harpe electrique*' had only just been published in 2023, by then the Spanish had already published several field studies.

4.3 - Researchers and Research

The researchers are usually entomologists or ecologists and not beekeepers. If the scientists labelled beekeepers 'unscientific and anecdotal' in their claims of colony loss, a label that was bound to cause offence; it was an accusation that could also be levelled at some of the subsequent research.

The term unscientific needs further explanation. In the second edition of her handbook on the Asian hornet, Dr. Sarah Bunker explains that 'in science significance is a standard test to see whether results could be down to chance'. The test of whether something is 'scientific' seems to be whether it has been subject to peer review and published in a scientific journal.

There is a very small group of scientists dominating this research in France and the issue is whether their over-reliance on modelling has become 'unscientific' in itself. Is it good science to use questionable data, run thousands of times through a model or multiple models, that may have been built for another purpose, until the scientist has a range of results that seem to agree? There is evidence of this over-reliance in the research on the hornet's effect on predation (see Section 8) and in the estimation of the potential cost of the hornet in terms of colony loss (see below).

Far too often findings were drawn on scant information and with only a theoretical understanding of the honey bee colony. However, this is unlikely to diminish the scientists' stated aspiration to advise beekeepers on managing the pest. There are notable exceptions to this, INRAE for example, who have their own apiaries from which to gather data, conducted useful early studies but then the research baton passed to a newly-founded ITSAP (French Institute for the Study Of Bees and Pollinisation) and it seemed to go very quiet.

> **GDSA 30 (Gard Department Bee Health Association) – January 2021 Newsletter –**
>
> (After almost 14 years of the hornet's presence in the department .....)
>
> *'Some (beekeepers) are upset and seem to discover that the Asian hornet is a real problem for beekeeping. It is not enough to raise alerts or demand quick fixes from others. It is through our collective, coordinated, realistic and pragmatic action that we will be able to control or at least slow down the proliferation of this invasive species'.*

The scientists' reliance on the scientifically unproven 'competition theory' to explain away the number of queens caught in spring trapping, rather than use common-sense, points to an agenda and flies in the face of the evidence that came out of the spring trapping study (see Section 10 and Section 3.4 Urban or Rural). Poidatz and Thiery didn't find evidence of it in their research at the INRAE Bordeaux campus and Thiery describes the theory as anecdotal.

Researchers often rely heavily on computer modelling to make up for the lack of real field data. Multiple models are used to derive a finding. Extrapolation is built upon extrapolation. The inherent danger in the reliance on modelling can be seen in the 2023 paper 'Economic costs of the invasive Yellow-legged hornet on honey bees' (no.25 - see Further Reading) where the model combined 'large-scale field data, niche modelling techniques and agent-based models' to produce an estimated annual loss of French honey bee colonies, due to the hornet, of between 2.6% to 29.2 %', a very wide range indeed.

Quite apart from the questionable value of such a wide-ranging estimate, it was only the higher figure that appeared in the publicly accessible 'abstract' of the paper and so it was that which was picked up by many and taken as fact, thus making the headlines (see also inset box in Section 9.3 – The 80% Myth.). The lower figure of 2.6% was buried in the body of the paper. To access that, you either had to have an institutional licence (i.e., be a scientist) or, in my case as a mere beekeeper, pay heavily to access it. The higher figure may have served the authors' purpose to highlight the potential cost, and presenting the abstract in that way might better attract further research funding, but to do so was oblivious to the alarm it would cause.

The lack of real data is also proving a hindrance when it comes to control of the hornet, especially during the establishment phase, when the need for information to inform decision-makers is critical. But there is a disconnect between what the frontline knows is happening on the ground and the scientists' perception of what is happening based on very incomplete data and it is not a problem confined to France. It's also a problem in the UK (see inset in section 4.4 – 'A disconnection from the reality?').

In France there is no unified database of sightings or nests. The vast bulk of nest data (i.e., the location of the nests reported, size, height etc.) sits with the individual departmental nest destruction schemes but hasn't transferred across into the MNHN's national database.

That same 2023 paper modelled that there were estimated to be 1.08 nests per sq. km. across the whole of France i.e., over 500,000 nests (but other scientists believe the total could be nearer 1,000,000 nests). To obtain data for their computer modelling, the researchers resorted to selecting target areas, driving hundreds of kilometres along roads in each area (in the autumn when the leaves had dropped) looking for nests, and making local enquiries! Not, one would have thought, a very scientific way of obtaining nest data but at least it was an attempt to get real data.

Even where field studies do take place, one has to question whether those directing the research have their priorities right or even manage their resources well. For example; it was very frustrating that the field study on the *museliere* took four years (2013-2016), observed 44 colonies (22 with muzzle, 22 without) in apiaries widely spread across SW France. Given the urgency of the situation facing beekeepers, might it have been better to have concentrated on fewer locations and on getting the results out earlier than 2019?

They might also have changed *museliere*s when they realised that those that they were using were distorting the results. Rather than change *museliere*s, the researchers proposed a further study with a larger mesh *museliere,* which proposal was unlikely to satisfy beekeepers desperate for solutions.

4.4 - Beekeepers Under the Spotlight

Beekeepers have not been immune from the direct attention of the scientists. In 2019 a group of French scientists published a paper entitled 'Science communication is needed to inform risk perception and action of stakeholders' (no.23 in Further Reading). The 'stakeholders' were the beekeepers, whom the scientists complained were acting without the benefit of scientific advice and were implementing measures to protect their colonies with a distorted view of the actual risk.

A view that was gained from (I quote) '*a nationwide-stakeholder survey*' to record beekeepers '*risk observation, perception and personal action against the Asian hornet*' (this was carried out in 2013 and so was prior to the publication of the 2019 paper on predation colony loss).

The beekeeper answering the questionnaire was heavily constrained with simple Yes/No or numeric answer boxes e.g., have you seen any hornet nests within 500m of your apiary Y/N? Same for hornets predating your hives – Y/N? How many dead colonies – number? Number of dead colonies presumed to be due to the hornet's predation? Were they using traps – Y/N? If so – how many – number? Commercial or homemade trap-Y/N? Commercial or homemade bait – Y/N?. No opportunity to say how many nests or hornets or how selective the traps were. Furthermore, 500m was well below the predation range of the hornet.

The question on the number of colonies presumed to be lost due to the Asian hornet, was a question to which the beekeepers couldn't know the answer (colony loss is multifactorial) and so the scientists were inviting them to guess. If the beekeeper had seen no nests within 500m but had 4 nests within a 1km predation range, had seen hornets predating their hives and said all hives were lost due to the hornet; how could the researcher know whether the beekeeper was spot on in their risk perception or not? The questionnaire was answered by just 419 beekeepers (0.6% of the 60,000 registered beekeepers), with 401 useable responses, which compares poorly with the 17,383 beekeeper responses to the ESA's annual winter mortality survey.

At that low percentage of response, statistical self-selecting bias was inevitable. Furthermore, the assessment of 'real risk' as opposed to 'perceived risk' was based on the MNHN's own database that held fewer nest records in total than are destroyed in one year in Brittany and a calculation of losses using computer modelling that depended on the scientists' own assumptions.

What was purported in the abstract to be a study of a much wider issue of science communication, turned into a focused and prolonged critique of the beekeepers' actions judged against, in many cases, the scientists' own science. The paper concluded that 'the results of this study highlights the need to improve the quality and quantity of risk communications between science and action (by stakeholders) in the early-stages of management plans'.

The release of the paper followed the publication of the paper into the effect of the hornet's predation. A paper that had shown that the beekeepers' concerns about the hornet had been justified all along. The MNHN must have known that the results of the ITSAP spring trapping study were about to be published (the MNHN and ITSAP had worked together on the study) and that it would confirm the beekeepers' contention that wide-area trapping worked to reduce nest numbers.

What appears to be an attempt to get in before the ITSAP technical note was released, would be followed in 2021 by yet another study that had been conducted years earlier (2008 – 2010) by MNHN and CNRS researchers; 'Not just honey bees: predatory habits of *Vespa velutina* (Hymenoptera: Vespidae) in France' (no. 61 – see Further Reading), which had established that a single hornet nest consumed around 11kg of other insects (circa 100,000 honeybee-sized prey) in a single season. The scientists' point being that if the hornet's predation was affecting honey bee colonies, bees were not the only insect affected.

It was as though, having been proved wrong about predation and about to be proved wrong on wide-area spring trapping, the scientists wanted to discredit the beekeepers using whatever study lay to hand; which was the point made by Christian Pons.

In other circumstances the science communication paper might have been laughed out of court but it was very damaging. Few in authority, those who would be influenced by it, would actually read the whole paper and see how shallow it had been but everyone would know what the real subject of the paper was and who the authors were talking about.

A disconnection from the reality?

A 2024 paper from the UK Centre for Ecology & Hydrology (no.22 in Further Reading) proposed modelling techniques to fill the gap in knowledge that hinders decision-makers in the early stages of an invasive pest's establishment. It wasn't the first paper to attempt this (see Further Reading for Section 3) and the authors had selected the Asian hornet for their modelling.

The model used 'worldwide Asian hornet occurrence data from 1993 – 2020', thus missing the 2022 surge in western Europe, and the records used from France were only for 2004 – 2014, thus also missing a surge in nest numbers in France in 2016 and 2018. The records from Spain were from 2010 – 2016. In Galicia alone, some 7,000 – 11,000 nests had been reported in 2016 but by 2018 there were reports of some 27,705 — 43,000 nests. An increase of that order in just two years would surely be very significant.

Careful filtering of the data that was used in their modelling, to avoid duplication, left them with just 23,000 records to enter into their model, which is fewer than the number of nests destroyed in Brittany alone in one year, never mind Galicia. One has to wonder about the value of such research when it is based on data that is so obviously incomplete.

Furthermore, the hornet nest's development is affected by the weather at critical times in its seasonal cycle but the model used annualised data. They were therefore looking at climatic favourability and not at the actual weather during the year at those critical times, thus missing what all the department schemes are convinced is the most important factor affecting nest numbers.

And no-one has explained the disproportionately high number of queens caught in spring trapping, other than to point to competition between queens, for which there is no scientific evidence but which points to a very large number of unreported nests.

The authors adopted the idea that the hornet favours areas disturbed by human activity and programmed that into the model i.e., the French idea that the hornet is an urban dweller. Although that may be true to some extent, it is questionable, so should any value for that factor be put into a model?

It is to no-one's benefit when the relationship between scientists and stakeholders degenerates to the level of acrimony. The beekeepers felt undervalued and not listened to, even though they were right. The scientists thought they had a wider view and responsibility that bordered on hubris but I wonder if they stopped to think where this would all lead. The scientists lacked data and the stakeholders' (beekeepers) cooperation was needed to get it. How likely is it that beekeepers will want to fill in a survey or cooperate with the MNHN or ITSAP in the future?

Beekeepers found themselves increasingly under the spotlight and heavily constrained by the ecologists, fearful to do anything that might upset the government, partly because they hoped for state aid to fund control of the hornet and for compensation to cover losses from the hornet's predation. As the GDSA newsletter in the inset panel suggests, many beekeepers were also waiting for the scientists to come up with a quick fix.

4.5 - Solutions to the Asian Hornet

There have been a number solutions to the Asian hornet put forward by scientists in France.

Around 2014 there was a flurry of suggestions that the hornet's limited genetic diversity would result in a sudden collapse of the hornet population and much was made of the presence in nests of diploid males and triploid females. A defect due to a lack of genetic diversity? Unfortunately, this phenomenon seems to be fairly normal in some vespids (no's 15 & 16 in Further Reading). As Eric Darrouzet (lead author of the 2015 paper, Production of Early Diploid Males by European Colonies of the Invasive Hornet *Vespa velutina* nigrithorax (no.15 in Further Reading) acknowledges ' *This indicates that the yellow-legged hornet can establish successful populations, even from a limited number of foundresses with low genetic variability'*.

Hornet numbers continue to increase and ten years on there is no sign of an H.G.Wells 'War of the Worlds' scenario where the nests burst open to reveal dying hornets. Safer for beekeepers to assume it is not going to happen than mark time waiting for it.

The following solutions have all been put forward by scientists worried about the side effect of indiscriminate trapping:

1. Fungi and nematodes have been suggested but nematodes would not affect enough hornets to make any meaningful difference to nest numbers and other scientists are understandably reluctant to see the introduction of something over which they may not have any control. Some still think that fungi might have a role to play (no. 21 – see Further Reading).
2. Saracenia pitcher plants are often mentioned by the MNHN and are still featured on their website. They are very attractive to the hornet but the maximum number of insects found in a plant was 7 being slowly digested. You'd have to plant an awful lot in the apiary to make

    a difference and of course they are not selective, so I am surprised that the MNHN even mention it.

3. European honey buzzards have been observed to attack the hornets' nests (no.18 in Further Reading) but here in the UK we currently only have about 40 breeding pairs, living mostly around the coast.

The current favourite is a pheromone trap that attracts the male hornets. The theory is fine, they have isolated a pheromone to attract the male sexuals in autumn, but they are not sure where mating takes place and so where to put the traps. They have also encountered problems with the trap itself and it is untested in the field on any practical scale — so they are not as optimistic as they perhaps were.

Anyone interested in this avenue of research should study the use of pheromone traps in the fruit industry. The design of the trap has to be matched to the target insect, a lot of traps are required to cover a small area, the pheromone used has to be stored under specific conditions because it degrades quickly. In short, potentially a very expensive solution.

There are 'Asian hornet' pheromone traps available online but I urge great caution if you are thinking of buying them. Especially if the trap arrives in a plain box with no instructions and a tiny glass test tube containing 4 or 5 rubber plugs that are reputedly laced with the chemical pheromone (i.e., not hermetically sealed in a foil container to be kept refrigerated) and the supplier turns out to be operating from a private address on a housing estate and not a business address.

Incidentally, the traps being mooted only use the pheromone plug in a cage to get the hornet to the trap, it still has to use a sweet bait to get them to go inside, that's not a selective trap at all.

These are all solutions conjured up by the entomologists and ecologists to provide a promised alternative to trapping but that is all they are, promises that years later are unfulfilled. The problem for beekeepers is that it is the scientific lobby that is driving the response to the Asian hornet, and when considering defences against the hornet the beekeeper can ill afford to wait.

4.6 Summary — Science and Solutions

The situation in France suffered from the absence of a government strategy. In the consequent vacuum, the public's response, with widespread non-selective trapping, threatened biodiversity thus causing environmentalists and ecologists to react accordingly. Beekeepers found themselves caught in the middle, unsupported and constrained in their efforts to protect their honey bee livestock. That must not be allowed to happen in the UK.

The scientific response was severely hampered by a lack of direction, funding and resource. Research took too long and was (and still is) too reliant on computer modelling in the absence of actual data. Some of that missing data was actually available but lay unused, presumably because it wasn't deemed by the scientists to be 'scientific enough'. Most damning of all are the field trials that didn't open hives or use other methods to find out what was actually happening inside the hives that they were supposed to be studying.

For a variety of reasons, the scientific response to the Asian hornet in France has been inadequate from the outset. Those scientific bodies that did their best for beekeepers (most notably INRA) found themselves on the side-lines and the government policy of inaction prevailed. Much of the science concentrated on predicting the spread of the hornet and its potential damage, expressed in monetary terms, in an attempt to get funding for more research.

The scientists labelled beekeepers 'unscientific and anecdotal', seemingly oblivious to the insult and unable to appreciate the value of the beekeepers' knowledge and experience. The release of papers, the purpose and timing of which seemed to be to discredit the beekeepers, is what Christian Pons was referring to in his editorial. The accusation of incompetence levelled against some of the scientists in L'Abeille shows just how acrimonious the situation has become.

If we are to avoid what has happened in France we should look more closely at the Spanish response, which has been much more collaborative, quicker to respond and quicker to get results published. Just as importantly, the Spanish have not been too ready to accept published research and have challenged it across a range of scientific disciplines.

The message for beekeepers in the UK, based on what has happened in France, is not to expect a sudden rush to research and certainly no quick results. In other words, don't wait for the quick fix it's not coming anytime soon.

Furthermore, whilst there are scientists that are undoubtedly on the beekeeper's side, it would be wise to recognise that, as Christian Pons alluded in his editorial, there are those scientists who are not actually on your side at all.

5 - Nest Destruction & Wide-area Spring Trapping – A Job for Beekeepers?

5.0 — Introduction

Nest destruction is an essential part of an integrated control strategy. Without it, nest densities can rise to intolerable levels that pose a threat to pollinisation (not just through the loss of honey bees), biodiversity and human health; quite apart from the hornet's effect on daily life. That does not make the Asian hornet's nests the responsibility of the beekeeper, any more than the beekeeper is responsible for wasp or European hornet nests.

In the pre-establishment phase, when eradication may still be possible, nest destruction is clearly the priority but the beekeeper should be in no doubt that nest destruction itself is a job for the professional pest controller. It should only be done by those trained to do it, using authorised destruction methods and only after a careful risk assessment.

Nest Destruction – not a job for amateurs!

A beekeeper from France turned up at a conference in the UK with a nest he had destroyed, in order to show his audience an Asian hornet nest. Fortunately, the conference was also attended by an NBU inspector who pointed out that importing such a nest was not allowed and promptly confiscated it before he could get it out of its bag. As the inspector and I examined the seized nest afterwards, one of the supposedly dead hornets started to come to life in the warmth of the inspector's hand!

What I would like to concentrate on in this section is whether nest destruction (with or without wide-area spring trapping) works as a longer-term control strategy when the hornet is established, and to give the beekeeper an appreciation of what is involved in terms of cost and effort. Furthermore, nest destruction may not be the solution some think it is and I question whether the beekeeper would not be better off concentrating their efforts elsewhere.

Even where the hornet is established there are those that are convinced that nest destruction is the only way to counter the Asian hornet and they dismiss apiary defences as ineffective. Such a view is mistaken:

1. Nest destruction can never be 100% successful because you will never find all the nests. Even in French departments with efficient schemes there are municipalities/communes that do not join the scheme and there are always unreported nests, especially in the countryside (see Section 10 – The Trapping Debate).

2. An early study by researchers from INRA and the CNRS (Centre National de la Recherche Scientifique) at the University of Toulouse, concluded that destroying 60% of nests = a 17% reduction in spread & 29% reduction in density. Even if an unrealistic 95% of nests were destroyed it would only result in a 43% reduction in spread & 53% reduction in density.

3. Nest destruction can give an illusion of control. Whilst it does have an effect, a much greater effect comes from the weather
(see Section 7.4 – Nest Development).

4. Apiary defences can be very effective if deployed correctly.

It is not possible to find every nest hidden in the countryside before the emergence of sexuals and so nest destruction cannot be a solution in itself. Photograph reproduced under licence - Retro Nico stock.Adobe.com

There have been many proposals to solve the nest identification and location problem:

- ▸ Heat sensitive cameras did not initially prove very successful because the nests were well-insulated and hidden in foliage so the camera had to be close to detect the nest. Improvements in technology may now make this an option but cameras with the requisite sensitivity and the drones to carry them are expensive, not to mention the need for highly skilled operators.

- ▸ Radar tracking of tagged hornets was another option. Tags were cheap but the radar unit very expensive.

- ▸ Peter Kennedy from Exeter University successfully demonstrated that using radio-telemetry to track radio-tagged hornets could dramatically reduce the time taken to find nests using equipment that was more affordable. The problem there seems to be that transmitter tags are too expensive.

- ▸ A similar technique developed by a team at the University of Nice Sophia Antipolis, at about the same time, yielded similar results but there too nothing has been adopted for wider use in the field.

The difficulty with these nest location techniques always comes down to cost and the availability of skilled operators versus the sheer scale of the problem. One would have thought they were ideal in the pre-establishment phase yet even our own NBU prefers to use the simpler track & trace techniques developed on the island of Jersey, even if they do take longer.

We'll now look at three well-organised department-wide nest destruction schemes, the first to see what it can achieve, the second to see what it costs in terms of money and resource and the third, to see the conclusion reached by one department (Ille-et-Vilaine) after 10 years in operation.

5.1 – Does Nest Destruction Work?

The Asian hornet arrived in the Brittany department of Morbihan in 2011, a department that, at c.6,000 sq. km., has roughly the area of Devon. A departmental nest destruction scheme started in 2015 as part of a 5-year study across a number of departments. It was very well organised and incorporated a wide-area spring trapping campaign. The rate of increase in reported nests that had been seen prior to the start of the scheme was slowed and in the years that followed, Morbihan reported some 35-40%

fewer nests than adjacent departments of similar size that also had nest destruction schemes.

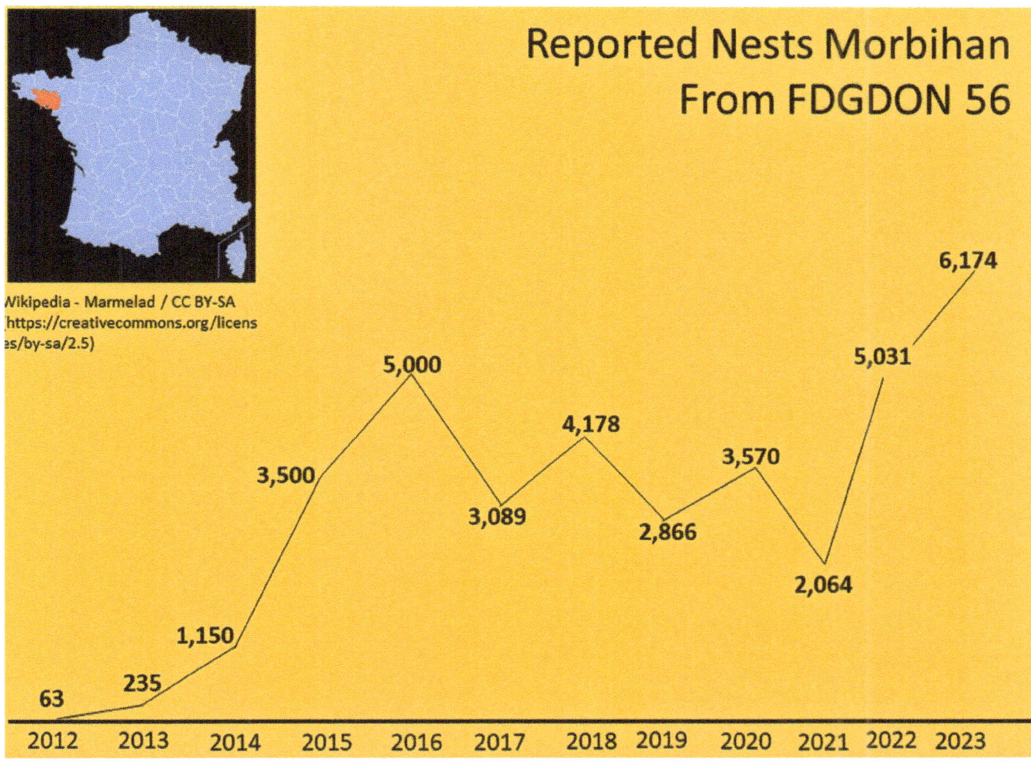

Reported nest destruction and spring trapping in the department of Morbihan 2012-2023

In setting up such a scheme it is essential to get beekeepers and local authorities working closely together. The Morbihan scheme was greatly aided by the fact that Giles Lanio, beekeeper and President of the Morbihan Beekeepers' Union (elected President of UNAF in 2015) and now an acknowledged expert on the Frelon Asiatique, was working closely with Joël Labbé, the French Senator for Morbihan. The latter known for his close connection with agriculture and beekeeping, and famous for regulating the use of phytosanitary products on the national territory, known as the 'Labbé law'. Giles mobilised the beekeepers and Joël mobilised the local authorities.

Nest destruction in Morbihan didn't solve the problem but it brought a degree of control. The year-on-year fluctuations in Morbihan's figures were, with the exception of 2023, all attributable to the weather. But the massive effort in cost and resource to maintain the scheme year after year is proving unsustainable. The increase in reported nests in 2023 was attributed by the FDGDON to a warm spring but the increase was not reflected to the same extent in adjacent departments and was likely due in part to the fact that some municipalities had withdrawn funding.

5.2 – What is involved in Effort & Cost – is it Sustainable?

I described the organisation of another nest destruction scheme, in the French department of Calvados, in my article 'Asian Hornet – The Window of Opportunity is closing' (July 2021 issue of the BBKA News).

Anyone contemplating a scheme in the United Kingdom needs to understand the organisation and complexity of such a departmental scheme. The beekeeper thinking of becoming a 'referent' should look carefully at the terms of reference for the role.

The first requirement is for a coordinating committee, which in this country would incorporate representation from the county council, emergency services, Defra, beekeeping associations, naturalists etc. In France the core costs of operating the scheme i.e., excluding the actual costs of nest destruction, are met by funding from a Council of Communes. I am unsure what the equivalent would be in this country. Such costs include communication, monitoring, reporting and the appointment of the Scheme Managers. This latter role is absolutely key to the success of the scheme.

The Scheme Managers have a number of duties:

- ▶ Communication with individual communes (parishes) to encourage them to sign up to the collective scheme. This involves agreement by the commune to fund their element of nest destruction and to undertake to report Asian hornet nests as part of the scheme's operation.

- ▶ Reporting on recruitment to the scheme, monitoring contributions and of course collating statistics on the operation of the scheme and the hornet's presence across the county.

- ▶ Recruiting pest controllers to the scheme. This may be as simple as getting prices so that landowners can choose a pest controller from a list or it may require pest controllers to sign up to an agreed tariff. Which it is will be determined by market forces.

- ▶ Training pest controllers on the risk from the hornet, approved pesticides and the protocols to be followed to ensure public safety.

- Carrying out an audit to ensure that pest controllers comply with the conditions of the scheme. Follow-up action in the event on non-compliance.

- Recruitment and training of 'referents' who are tasked with investigating nest reports to confirm that the pest is the Asian hornet (typically 20% of reports in France are for wasps and European hornet nests).

- Liaising with the local commune (parish) clerk, who could be the initial recipient of nest reports from the public.

- Manning a reporting hotline and on-line reporting system, tasking referents, managing the billing process for pest controllers.

- Operating an on-line geo-referenced database to log nest reports, the precise location of the nests, authorisation of destruction and subsequent actions.
  The cost of the database is part of the core cost of the scheme.

- Communicating with the public and operating a departmental-wide (county-wide) publicity scheme.

FGDON Ille & Vilaine
Fédération des Groupements de Défense contre les Organismes Nuisibles d'Ille et Vilaine

NE CONFONDEZ PAS LE FRELON ASIATIQUE 25-30 mm

AVEC la Scolie des jardins
Pacifique, piqûre insignifiante en défense ultime 35-40 mm

AVEC le Frelon européen
piqûre potentiellement dangereuse 30-40 mm

la Sesie vespiforme
Papillon à l'aspect frelon innofensif 20-27 mm

le Syrphe volucelle
Innofensif 20 mm

le Sirex géant
Innofensif 40 mm

Plus d'information actualisée sur www.fgdon35.fr

FDGDON Poster – Misidentification is a problem

Another key role is that of the 'Referent' and this is a role often undertaken by beekeepers on a voluntary basis. The referent's role is to:

- Receive a report of a nest either from the Scheme manager, commune (parish), the emergency services or sometimes direct from the landowner. In the latter cases ensuring that the report is logged into the database.

- Investigate the nest report and confirm to the Scheme Managers that nest destruction is required under the scheme.

- Assist the Scheme Managers with the follow-up audit process.

- Liaise with beekeepers and assist in the education of beekeepers.

- ▸ Liaise with the public and give briefings at village halls etc.
- ▸ Report to the Scheme Managers on the hornet's presence in their commune and assist in the compilation of statistics.
- ▸ Where area-wide selective spring trapping is in operation, to assist in its organisation, monitoring to ensure compliance, and collation of statistics to Scheme Managers.

A pest controller who signs up to the scheme is required to use only approved pesticides (e.g., Calvados authorises only Pyrethrum) and to ensure nest destruction is carried out to the scheme protocol. The Scheme Managers contact a percentage of those who have had nests destroyed to confirm compliance with the protocol and in the event of non-compliance the scheme will not pay the pest controller, who may be removed from the scheme.

In France, funding of nest destruction itself was typically split a third to the department, a third to the commune and a third to the landowner on whose land the nest is located. In departments where such schemes operate the sign-up rate is over 95% of communes. This very high rate is down to two factors:

1. The cost of a nest's destruction is typically 120 euros. The 1/3rd cost to the individual is acceptable and there are obvious savings in a scheme with agreed tariffs.

2. The alarm caused by the presence of nests in the community results in a ready support from the public for the scheme.

Here are the bare statistics for Calvados (5,535 sq. km.) in 2019. FDGDON appointed to manage scheme:

- ▸ **427** member municipalities (communes) signed up
- ▸ **58** pest controllers in tariff scheme
- ▸ **234** new referents trained in 2019
- ▸ **1544** nests destroyed in 2019 = 146,680€

The above statistics were reported in 'BILAN TECHNIQUE DU PROGRAMME DE *LUTTE COLLECTIVE* CONTRE LE FRELON ASIATIQUE - DEPARTEMENT DU CALVADOS – ANNEE 2019'. The cost of scheme's management was not given.

However, 2019 was a quiet year for Calvados. In 2021 the department had almost 3,000 nests to deal with. Even the 314 referents in 2021 were not enough to have a referent in every commune or municipality. The 45 pest controllers destroyed nearly 3000 nests at a cost of around 120 euros per nest in destruction fees alone and that is recurrent expenditure. If the landowners paid a third of that cost, the local authorities still had to find 240,000 euros plus the cost of running the scheme. Incidentally, that number of reported nests is on a par with what was reported in Flanders in 2023.

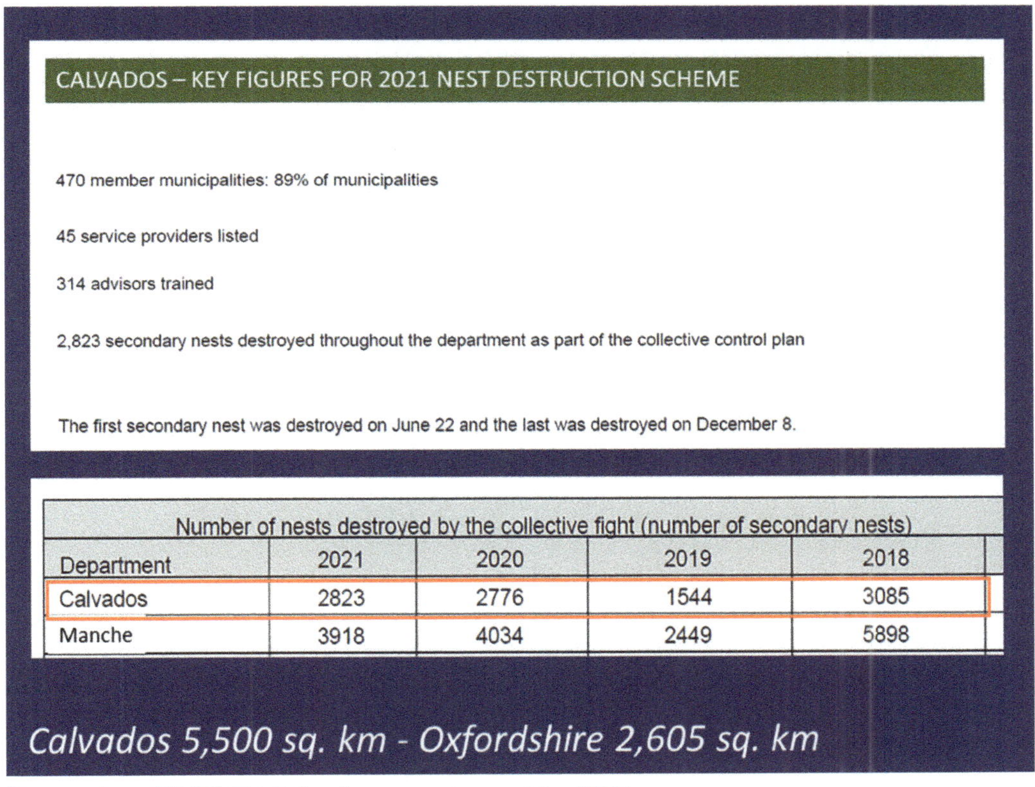

CALVADOS – KEY FIGURES FOR 2021 NEST DESTRUCTION SCHEME

470 member municipalities: 89% of municipalities

45 service providers listed

314 advisors trained

2,823 secondary nests destroyed throughout the department as part of the collective control plan

The first secondary nest was destroyed on June 22 and the last was destroyed on December 8.

Number of nests destroyed by the collective fight (number of secondary nests)				
Department	2021	2020	2019	2018
Calvados	2823	2776	1544	3085
Manche	3918	4034	2449	5898

Calvados 5,500 sq. km - Oxfordshire 2,605 sq. km

Extract from FDGDON Calvados annual report for 2021

5.3 - Ille-et-Vilaine, Brittany – The Conclusion after Many Years

Ille-et-Vilaine is the eastern-most department of the four Brittany departments. In 2023 it destroyed 8,722 nests as part of its department-wide nest destruction scheme. The scheme had started in 2014 and for a number of years had included wide-area spring trapping of foundress queens.

Ille-et-Vilaine abandoned spring trapping for a number of reasons that we'll look at in Section 10 – The Spring Trapping debate. As far as nest destruction is concerned the FDGDON, the scheme managers, concluded that although nest destruction is not a perfect solution it did considerably reduce the pressure of the hornet on the environment and the early detection of nests limited the risk of attacks. If it were not done the environment would be saturated with Asian hornets from September to mid-November with all the impacts that held.

Their final conclusion was (I quote the FDGDON 35 report):

'In Ille-et-Vilaine, the 'one nest discovered = one nest destroyed' policy is made possible thanks to the exceptional mobilization of local authorities (Departmental Council, communities of municipalities and municipalities) which provide all or part of the local costs of destruction. When the destruction remains the responsibility of individuals, only 2 out of 10 nests are treated. Ideally, early detection of nests should be used to improve overall effectiveness'.

The hornet surge year of 2022 saw a huge increase in hornet nests across France. In many departments nest numbers doubled. It was clear that the cost was unsustainable but at the same time public concern was demanding action.

5.4 - The French National Fight Against the Asian Hornet (2023)

Following the surge in 2022 and the increasing pressure for a national control of the hornet, it was decided to implement a 'National Fight Against the Asian Hornet' to be implemented in all departments.

Masterminded by GDS France, the over-arching animal health body, the plan's weakness was that it was a plan on paper with no funding attached and a very poor substitute for the publicly subsidised schemes. It relied almost entirely on volunteer effort focused at department level on the bee health associations and beekeeping associations. It was entirely reliant on the local schemes being able to raise funding and to recruit the volunteer referents at commune level.

The referents were expected not only to carry out the duties of a referent but because there is a dominant scientific representation at the top level, referents were also expected to populate databases with additional information and they were expected to analyse the contents of traps every week. Those requirements are at the heart of the disaffection amongst beekeepers with the collective schemes and they form the bulk of the referents.

There was an absence of funding for nest destruction, which before the publicly subsidised '*lutte collective*' came into operation, meant that nests were often not destroyed (far less than 20%). Furthermore, it was very worrying that the new scheme raised the spectre of the French joining the Belgians in only destroying nests located below 5 metres in height because only nests below that height are deemed to be a threat to the public.

The new scheme got off to a shaky start and a report by FNOSAD in September 2023 on the first year of operation found that:

▸ The scheme had only been adopted in 40% of departments.

▸ Only 35 departments had a steering committee or departmental referent.

▸ Only 42% of the schemes had operated spring trapping.

▸ 35% of GDSA's were doing the nest destruction themselves, which suggests they hadn't raised the funding required.

An article on the scheme in L'Abeille, reviewing the report on the first year, was headlined 'Lutte Contre le Frelon Asiatique – La déception', which means the disappointment. I prefer the word deception because it was never going to work.

5.5 – National Plan 2024 – A Plan with no Clothes?

In the 2024 version of the French National Fight Against the Asian Hornet, now with the snappy title 'Stratégie et plan national de lutte contre le frelon asiatique à pattes jaunes'; spring trapping was only envisaged in areas with medium predation (i.e., 1-3 hornets always predating in front of the hive). As nest numbers grew, nest destruction was confined to nests within a 1km radius of apiaries with heavy predation (i.e., 4> hornets always predating in front of the hive). An emphasis was to be placed on beekeepers implementing measures in the apiary. At the time this book was going to print, I had had no update on how it had gone.

But it is a plan with no substance, tied up in tiers of bureaucracy, far too dependent on beekeeper volunteers and on funding by local authorities that are reeling from other pressures. Furthermore, the restriction on spring trapping is an acceptance that nest destruction is not feasible once the hornet is established.

The new plan is almost impossible to manage and will lose all support for funding when the public understand that it is an illusion of action, a Hans-Christian Anderson's 'The Emperor's New Clothes', and that they have to pay the full cost of nest destruction on their properties.

A hive under heavy predation and on the point of total foraging paralysis. There are five hornets hawking the hive which should have been fitted with a *museliere* and *harpes electrique* deployed. Photograph reproduced under licence – Rodolphe stock.adobe.com

5.6 - Beekeeper Nest Destruction Teams in Wallonia

In Belgium the responsibility for nest destruction has passed to regional/provincial authorities and, except where there is an imminent threat to public safety or public land is involved, nest destruction is a cost for the landowner. However, individual provinces seem to be liberal in their interpretation on what is an imminent threat, one even extended that category to a nest inside a house or courtyard. Some fire brigades will destroy a nest for a flat fee that is cheap at 100 euros. The same thing happened in France prior to the change in 2012 that allowed the *lutte collectives*.

Schemes have been set up in Wallonia to have beekeeper nest destruction teams responding to beekeeper requests to destroy Asian hornet nests within 1km of their apiary. Eighteen teams were set up in 2023, provided with the necessary equipment and trained in its use. Each team covers an area of approximately 900 sq. km. or just under a third of the area of Cambridgeshire. A single team is therefore covering the equivalent of one of that county's district councils.

If Beekeeper Nest Destruction Teams covered all of Belgium…

▸ Nest destruction within 1 km. of an apiary is at the request of the beekeeper whose apiary is affected.

▸ A 1 km. radius gives a circle with an area of c.3 sq. km.

▸ There are 10,000 registered beekeepers in Belgium

▸ Belgium has an area of c.30,000 sq. km.

Do the maths …

Even though the scheme only got started towards the end of 2023, the Minister of Agriculture (Willy Borsus) had already published proposals to double the number of teams for 2024 and was considering plans to meet at least some of the cost of running them. Those proposals were put on hold pending elections and at a time when the Wallonia nest reporting database appeared to show a marked reduction in nest reports.

Unfortunately, the databases for Flanders and Wallonia were no longer giving an accurate picture of the situation following the devolvement of nest destruction to regions and provinces. At the end of 2024 the reporting map for Flanders was still showing large areas where nests had been reported but action to deal with them had still not been confirmed. In Wallonia, the nest reporting was very patchy, some areas showing an increase in nest numbers and others hardly reporting any nests at all. The one thing that Belgium had that France did not, was a coherent picture of Asian hornet nests but now they don't even have that.

For how long the beekeepers will be able to keep up the effort remains to be seen, especially if the general public realise that there is an alternative to paying a pest controller; simply ask a beekeeper with an apiary within 1km to have the nest destroyed by a beekeeper team!

Destruction of an Asian hornet nest. Note the special suit; ordinary bee-suits are not suitable. Eye protection and thick gloves are essential. Photograph reproduced under licence – Lothar Lenz stock.adobe.com

5.7 · What is the Beekeeper's Role in Collective Action?

Beekeepers should not expect the current situation (as of 2024) to last forever. Our NBU has done a magnificent job in often trying circumstances. But our bee inspectors are needed in their primary role. Furthermore, it is not reasonable to expect those who have performed their duties with such fortitude to do so indefinitely.

The focus in this country in recent years has been very much concentrated on the Channel Island of Jersey and their track & trace is held up as an example to beekeeping associations. But Jersey is a small island of only 120 sq. km. Translate that to an English county 25 or even 50 times the area. Even in the Maritime North Climatic Zone, we are now talking about thousands of reported nests in an area the size of a county. Look at the table of resources required for Calvados; do we have any chance of getting that level of resource in manpower or monetary terms?

The '*lutte collectives*' mitigated the problem, but the simple fact is that without public subsidy far fewer nests get destroyed. It is not obvious who could undertake the scheme manager role, which I believe is beyond any county beekeeping association. No private landowner will want to pay the full cost of nest destruction, unless destruction is unavoidable, and I cannot see our cash-strapped county councils wanting to pay for intervention by the fire service. What then is the beekeeper to do?

In an editorial to his members, the President of the SNA (Frank Aletru) urged beekeepers not to adopt a fortress mentality when it comes to participation in collective schemes. The message was simple; no apiary stands alone and the only way to deal with the hornet is to fight it on all fronts, but he wasn't talking about track & trace or nest destruction (see inset below), he was referring to spring trapping.

'For years, we have advocated the establishment of a collective and selective fight against the Asian Hornet, before the arrival of spring. Then at the end of the season, strengthening the protection of colonies by putting in place effective muzzles and new generation harps.'

Frank Aletru – President of the SNA and Vice-President of the EFBA - 2024

In the pre-establishment phase AHAT's (Asian Hornet Action Teams) have a most important role to play in educating the public to report sightings and nests. Maintaining vigilant observation of bait stations is another valuable task that, whilst they are still in the field, can free up the NBU's limited resources.

Nevertheless, the beekeeper should consider very carefully where their primary duty as beekeeper lies and what effort is sustainable in terms of effort and cost to themselves. Join in collective action if you can but it is all too easy for this to evolve into tracking the hornets to their nests and even nest destruction itself.

I urge great caution with regard to both because it is a poisoned chalice. It will be all too easy for the beekeepers to be left holding the baby when the NBU has to withdraw and beekeepers will become a scapegoat for the public's frustration if nests are not destroyed. Furthermore, an association in the public mind between an invasive pest and beekeeping may not be one to foster.

There is the argument that it is better in the longer term for beekeepers to stand back once the NBU withdraws and let public pressure force local authorities and government to act, rather than attempt to run things themselves, thereby allowing those authorities to hide behind a semblance of action to which they are not contributing other than in an advisory role.

Furthermore, in track & trace the beekeeper is now walking towards a danger that they may not fully appreciate and at some risk to themselves (see section 6 - 'The Hornet and Humans').

Once the hornet is established, effort should be concentrated on the apiary and the area that surrounds it. This approach is at the core of the French National Plan 2024 and is centred on three actions, but the first is not a beekeeper action:

1. The destruction of nests within a 1-kilometre radius of an apiary that is suffering heavy predation.

2. The implementation of selective spring trapping by beekeepers, for a limited period, around apiaries that are experiencing medium/high level predation.

3. The use of defensive measures in the apiary itself.

5.8 – Summary on Nest Destruction.

It is clear that nest destruction is expensive, requires a lot of resource, but does not work as a control strategy once the hornet is established. It can only mitigate the problem. Where there is no public subsidy, far fewer nests get destroyed; that is significant when it comes to assessing the level of threat to the public but that is not a matter for the beekeeper.

All the evidence is that nest destruction destroys at best 50% of nests (see also Section 3.4 - Urban or Rural and Section 10 – The Spring Trapping Debate). The majority of nests are not detected until late summer or beyond, by which stage the damage to biodiversity has been done. The emphasis in France is currently on earlier reporting, location and destruction of nests and away from spring trapping but unless reporting, location and nest destruction can achieve the destruction of an unrealistically high percentage of nests, it will be a never-ending task and an unachievable goal. It is not sustainable.

Where wide-area spring trapping of foundress queens has taken place in conjunction with nest destruction, three points have emerged:

1. Even with an efficient scheme, a vast number of queens still survive hibernation to emerge in the spring (see Section 10).
2. There is a substantial reduction in nests compared to schemes that do not have organised wide-area spring trapping (confirmed by ITSAP).
3. Wide-area spring trapping, especially that involving the general public, results in an unacceptable level of non-selective trapping.

In short, wide-area spring trapping works to reduce nest numbers but is not controllable and therefore is not acceptable. The involvement of beekeepers in wide-area trapping is a moot point because they should be fully engaged in carrying out spring trapping in the vicinity of their apiaries, which trapping is acceptable.

There are those scientists that argue that spring trapping has not been proven to work whilst espousing nest destruction. I suggest that is it the benefit of nest destruction as a strategy that is not proven and that the question should be, is there a greater potential benefit in wide-area spring trapping, if it could be controlled? If that were the case, would it not be better to put the effort there?

6 - The Hornet and Humans – The Risk to Human Health

6.0 - Introduction

Beekeepers are going to come up against the Asian hornet, whether in the apiary or if they choose to get involved, as *'referents'* (those who confirm a nest is an Asian hornet nest), with track & trace, or nest destruction. The government's line that the hornet does not pose an increase in risk to human health is not supported by evidence from Spain or China. It is therefore a good idea to know what you are up against and the risk to yourself.

Public health advice is not within the scope of a beekeeper's guide to defence.

Nevertheless, the age demographic of beekeepers and the mere fact that you are a beekeeper puts you at increased risk, so I think we need to look at this more closely. I have to warn you that much of what follows contradicts the current official view and so I have included a very comprehensive reading list.

6.1 - Vespa Velutina – the Killer Hornet?

<u>Is the hornet aggressive to humans?</u> - Away from its nest, *V. velutina* is not known to be aggressive towards humans, this includes in the apiary, although it will sting if handled. Deaths are fortunately rare.

One situation to watch out for is when the hornets have overwhelmed a hive and are in the process of robbing it. There have been reports that the hornets will become aggressive in defending their prize. Best to let them get on with it, at least it will keep them busy and away from other hives.

Be aware that hornets can project an irritant that can be very painful if it gets into the eyes (see 39 – Ocular Lesions… in Further Reading). Early on in the invasion one French fireman fell off a ladder following an attack involving so-called 'projected venom'. Eye protection is important when dealing with the hornet, as was discovered years later in Jersey.

Apparently, that danger had not been known to them, although it was well known in France.

Nests were initially located high up in trees, which is probably why there were not many reports of serious envenomation incidents in humans in the first few years of the invasion (no.36 in Further Reading) but 12 years later it was a very different situation in Brittany where typically 50% of reported nests were to be found under 5 metres in height, and in one year it was 70%. That is important because it is within 5 metres of a nest that attack becomes a very real and serious risk. When the hornet can be very aggressive and will quickly mount a mass attack on an intruder.

Within 5 metres of a main nest there is a serious risk of a virulent mass attack by the hornets. Stay away.
Photograph reproduced under licence – Thomas Lenne stock.adobe.com

An astonishingly brave team of researchers, led by Dr. Moon Bu Choi of Kyungpook University in South Korea, set out to discover what the parameters were that excited a *Vespa velutina* nest to attack a human intruder and the best thing to do should you attract its attention (no's. 28 & 29 - see Further Reading). It is a shining example of dedication and perseverance at no small risk and the papers are well worth reading but the bottom line on safety is:

▸ Do not go near an Asian hornet nest.

▸ Do not under any circumstances try to deal with one yourself, that can be very dangerous.

▸ If you find you find yourself too close to a nest, move away without making rapid movements or vibration because those can trigger a virulent mass attack.

▸ If you find yourself under attack, run away as quickly as you can and get as far away as you can. You will take far fewer hornets with you if you run than if you walk, and standing still or squatting down will not help.

▸ Shield your face and neck with your arms and hands; stings to the arms and hands are much less likely to result in a rapid onset anaphylactic shock than if you are stung on the face or in a core body area.

▸ Wearing a broad brim hat results in fewer stings.

▸ Don't wear black or brown and if you are going out to a fancy dress party, take off the grizzly bear suit before going over to look at the buzzing noise in the hedge (you are beekeepers, you already know this).

<u>Is the hornet an increased risk to human health?</u> - There are several aspects to this; the overall risk of being stung, the risk of death, the risk of systemic reaction, and a toxic reaction to multiple stings. *V. velutina'*s venom is very complex and its effect is too readily dismissed. Whilst declared to be not more poisonous, per se, than that of our own hornet, there are reasons to challenge that view and there are aspects to the venom that are not fully appreciated. Furthermore, the risk of being stung is greater.

In 2021 the magazine Biology published a paper by Dr. Xesus Feas PhD (Academy of Veterinary Sciences of Galicia) titled 'Human Fatalities Caused by Hornet, Wasp and Bee Stings in Spain: Epidemiology at State and Sub-State Level from 1999 to 2018' (no. 30 - see Further Reading). The paper reported what on the face of it appeared to be a low number of deaths in Spain from hornet, wasp and bee stings (78 deaths over the 20 years from 1999 to 2018). Some took this to be a confirmation that the risk to human health was very low; but they should have read the whole paper because what Xesus Feas was actually highlighting was that there had been an increase in the rate of fatalities from hornet, wasp and bee stings in Galicia and Asturias, two of the most infested regions of Spain, following the establishment of the Asian hornet in those regions.

He warned that *'due to its habits, abundance and its broader distribution, the risk that Vespa velutina represented to human health is unmatched by other Hymenoptera native species'.* This goes to the core of the problem with the Asian hornet. Its aggression in defending its nest, the sheer number of nests once it gets established, and the fact that the nests are usually outside and are often hidden low down in areas of human population.

Dr. Feas had called on Spanish data for the mortality rate in the category X23 - hornet, wasp and bee sting fatalities. Looking at the data in his paper; in the 15 years prior to the hornet becoming established in Galicia (2014), the mortality rate in X23 had been less than 1 death for every million inhabitants. Significantly, I note that for 9 of those 15 years there had been no fatalities recorded at all in that category.

Following the hornet's establishment, the rate increased and in the hornet surge year of 2018, it tripled to 2.22 deaths per million inhabitants, which in a region of 2.7 million people was 6 deaths from stings in a single year. That is 4 more deaths than in any of the previous 19 years that had actually recorded a death in category X23 and that was in a population equivalent to that of Greater Manchester.

Dr. Feas concluded his paper by saying that *'In light of these findings, there is evidence …. to abandon the notion that the invasive species (Vespa velutina) poses no greater risk to human health than a bee'.*

2018 was a turning point in Galicia; the Galician Health Services implemented new measures and a specific quick route of care for people presenting at health centres or hospitals with the symptoms of a systemic reaction following a Hymenoptera sting, namely:

- ▶ Adrenaline kits were distributed to all health centres.
- ▶ A quick route of care for people exposed to insect venom and suffering a systemic reaction included:
    - Systemic Reaction Treatment
    - Specific analysis to improve the diagnosis

- Referral to an allergy clinic within 15 days

- Victims were issued with an adrenaline auto-injector pending their appointment at the allergy clinic.

This was a sensible and proportionate response by the Galician health authorities and a recognition that the hornet is a risk to human health that needs to be acknowledged. This is not the case in either France or the United Kingdom. So where did the UK government get the idea that there is no increased risk to health? In 2009, a report by Luc de Haro et al (no. 36 - see Further Reading) had concluded that in France there was no increased risk to health. But the report was based on data from the years 2004 – 2008, the first 5 years of the invasion, when nests were far fewer and usually high up in trees, where the public was not exposed to them.

In the 20 departments that Haro looked at, the hornet had been present in 11 for only two years and in 7 for only one year, and 'present' merely meant the first nest had been reported, which was far too early to judge. In a perverse twist, the paper mentioned that the risk was different in places like Indonesia where the hornet nested in or near human habitation; which is exactly the situation that exists in France now.

> In Indonesia *V. velutina* is apparently considered as an aggressive species. This reputation is linked to the fact that nests are frequently built in villages thus greatly enhancing the possibility of attack.
>
> 2010 D'Haro Report

There was an all-too-brief follow-up report in 2015 but it only looked at national data as a whole from 2009-2013 and so the results for the departments with the hornet were diluted amongst all the other departments. Nevertheless, it did note an increase in cases associated with the Asian hornet; although there was reference to a fashion for people to 'claim they were stung by an Asian Hornet', a rather unscientific explanation. Significantly, there doesn't appear to have been a study since the 2016, 2018 and 2022 surges.

But there is much more to this than just deaths and that is the number of people suffering a systemic reaction to the hornet's sting. Where that happens, the reaction can be rapid and is often severe. Commentators on the risk of the Asian hornet to human health fail to mention that a person can be sensitized as a result of a previous envenomation (i.e., sting) without being aware that the exposure has sensitized them. Where that has happened, it is the subsequent exposure that causes the reaction.

A link from *V. velutina* venom to other vespid venoms had been flagged by doctors in Navarra as early as 2014 (no.31 - see Further Reading). Normally, sensitization is specific to an insect but what clinicians are reporting is that *Vespa velutina's* venom has a similarity to other venoms that can trigger a systemic reaction in a sensitized person even if they had not been previously stung by it.

In 2021, a case study of 100 referrals to the Allergy Clinic at the regional University Hospital in Santiago de Compostela revealed that 77 were referrals as a result of *V. velutina* stings, concluding that *V. velutina* was the commonest cause of Hymenoptera anaphylaxis in that region. A very high percentage were first time reactions to a *V. velutina* sting thus confirming the Navarra warning. (no. 32 - see Further Reading: Anaphylaxis to *Vespa velutina nigrithorax* Vidal C et al.).

In a separate 2021 paper , Carmen Vidal, Professor of Medicine at the University of Santiago de Compostela and head of the allergy dept, confirmed the exact nature of the cross-link to not just other vespid venom but also to *Apis mellifera* venom (no. 35 - see Further Reading: The Asian wasp *Vespa velutina nigrithorax*: Entomological and allergological characteristics – Vidal Carmen). An Asian hornet sting could therefore trigger a reaction in someone sensitized by a bee sting.

A Chinese report (no. 33 – see Further Reading) that analysed the clinical outcomes of Asian hornet stings that had provoked a severe toxic reaction, had prompted an analysis of *Vespa velutina* venom in 2014 (no. 34 – see Further Reading). The sequencing showed that the venom is very complex with over 300 identified components, some of which are rarely found in Hymenoptera, one is more usually found in sea snake venom. The report concluded '*The Asian hornet is considered to be a dangerous invading predatory species, which has become a public health problem in China, and will likely become a public threat in its recently colonized countries*'.

The two aspects to the venom considered by clinicians to be the most dangerous are; its propensity to trigger an allergic response leading to a systemic reaction, which is likely to be severe (3-4 on the Mueller Anaphylaxis Scale). A quite separate issue is its toxicity, which can be a problem with multiple stings, especially in older people whose health may already be compromised. The Chinese studies indicate that the risk-of-toxicity threshold starts to increase over 10 stings.

<u>Are beekeepers immune to *V. velutina* venom?</u> Unfortunately, even if the beekeeper has been regularly stung and seemingly has built an immunity to *A. mellifera* venom, that does not guarantee immunity from the hornet's venom. There have been a number of cases in both France and Spain where beekeepers have died or have been hospitalised after a single Asian hornet sting. An NBU inspector had to be treated after suffering such a reaction after being stung, having disturbed a nest during track & trace operations. The cross reaction of *V. velutina* venom with *A. mellifera* venom, that may lead to a systemic reaction, does not therefore appear to mean that an immunity to A. *mellifera* venom guarantees an immunity to *V. velutina* venom.

Key points to understand when it comes to being stung by an Asian Hornet

- ▶ The hornet's venom can cause a systemic reaction in a person who has been previously sensitized by another vespid (e.g., wasp) or a honey bee.

- ▶ The person stung may be quite unaware that they have been sensitized by a previous sting; it is a subsequent sting that is the problem.

- ▶ The number of people who are sensitized is small but as the reaction to the Asian hornet's sting is often rapid, it is very important to recognise the symptoms of a systemic reaction and, where it is apparent that such a reaction is happening, get immediate help.

- ▶ Severe reaction symptoms may include: swelling in the throat and/or mouth • difficulty breathing • severe wheezing • feeling faint, dizzy, or very sleepy • itching and swelling away from the site of the sting • severe abdominal (stomach) pain.

- ▶ Multiple stings carry an additional toxicity risk. Seek medical assistance if you have been stung more than 10 or more times.

<u>Are beekeepers at increased risk from *V. velutina* venom?</u> Compared to the general population, beekeepers have a much higher lifetime risk of a systemic reaction through their exposure to honey bee venom, which

can leave them sensitized. Even if they are currently beekeepers they are still at increased risk (no's. 40 & 41 - see Further Reading: Papers by Carli T, Locatelli I, Košnik M, Kukec K). All beekeepers should be aware of the symptoms of a systemic reaction, not just because of any risk from *Vespa velutina* but just as importantly, from their own bees.

6.2 - Vespa Velutina – Victims of Stings

If we get away from the extreme examples (death and systemic reaction) and look at how many people are stung, it's quite common in areas where the hornet is established.

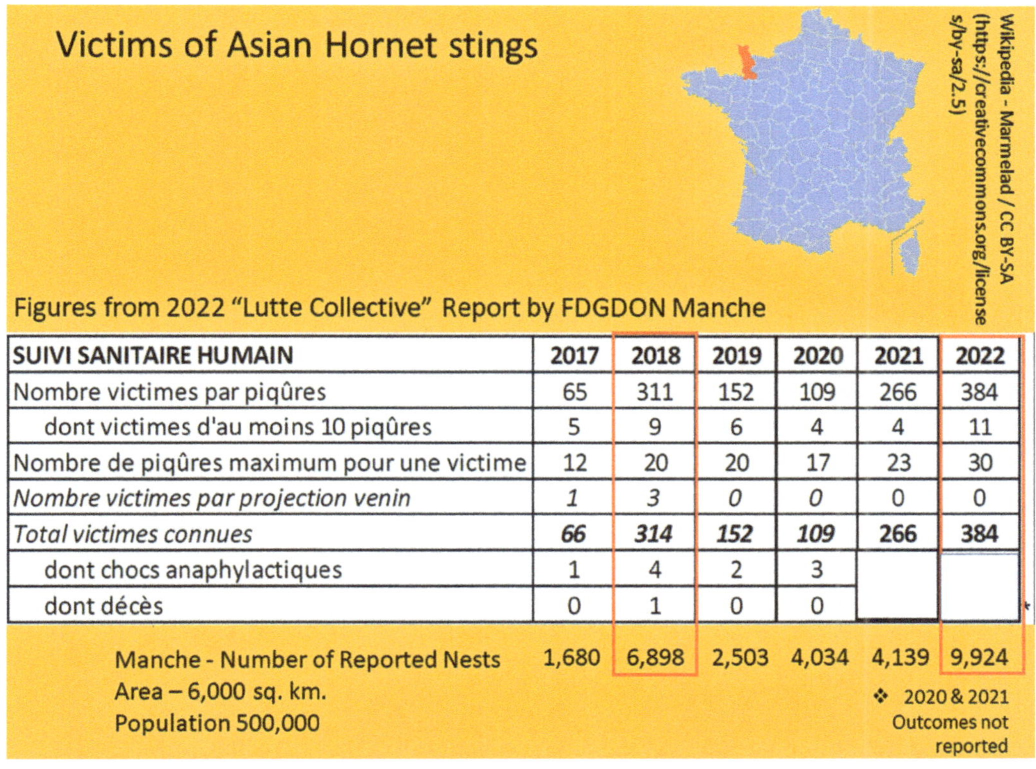

Figures from 2022 "Lutte Collective" Report by FDGDON Manche

SUIVI SANITAIRE HUMAIN	2017	2018	2019	2020	2021	2022
Nombre victimes par piqûres	65	311	152	109	266	384
dont victimes d'au moins 10 piqûres	5	9	6	4	4	11
Nombre de piqûres maximum pour une victime	12	20	20	17	23	30
Nombre victimes par projection venin	*1*	*3*	*0*	*0*	0	0
Total victimes connues	*66*	*314*	*152*	*109*	266	384
dont chocs anaphylactiques	1	4	2	3		
dont décès	0	1	0	0		

Manche - Number of Reported Nests	1,680	6,898	2,503	4,034	4,139	9,924

Area – 6,000 sq. km.
Population 500,000

❖ 2020 & 2021 Outcomes not reported

Table showing the number of victims of Asian hornet stings reported in the French department of Manche and the number of reported nests in those years.

The French department of Manche is c.6000 sq.km. (i.e., the area of Devon but with less than half the population). The table above shows (descending order), the number of victims, those with more than 10 stings, maximum number of stings for any one victim, number of victims of 'projected venom', total number of victims – those with anaphylactic shock or dead (but note that records of those with anaphylactic shock or dead ceased to be reported by the communes after 2020).

6.3 - A Duty of Care by Employers and Associations

It is important to keep the risk to human health in perspective. Deaths are very few and reports of 'referents' coming to grief in the course of their duties are also thankfully rare but they do happen. However, I question the idea that beekeepers, including bee inspectors, are best suited for duties that bring them into a situation where they are at increased risk by virtue of their own vulnerability and their exposure to attack in the course of track and trace or nest destruction (nos. 40 & 41 in Further Reading). It is akin to sending someone who may be genetically disposed to a cancer into a situation where a carcinogen for that cancer is known to be present.

Such exposure to the risk should only be contemplated if the participant is fully aware of their own susceptibility to a systemic reaction. Furthermore, any of those who are in an additional risk category through age or other health factor, should be excluded from such duties. It may be through ignorance, indifference, or exigency but the current attitude to this risk is to my thinking too casual.

I am mindful of the case in 2023 when Olivier Quesnot, a former mayor of Tilly-sur-Seulles in the department of Calvados, died as a result of a mass attack by Asian hornets whilst acting as an AHAT referent. I did not know Olivier but I had been to Tilly-sur-Seulles on a number of occasions. It was a tragic end to the life of someone who had served his community and his department for so many years. Had Olivier known of the toxicity risk from multiple stings, might he have thought twice before volunteering to be a AHAT referent, given it was later reported that he had a heart problem?

Employers or associations that send individuals or teams into the field (volunteers or not) have a responsibility and duty of care to ensure a proper risk assessment is carried out and to make proper provision for dealing with any incidents that occur.

Furthermore, an auto-inject adrenaline pen is not a guarantee of safety. There have been a number of deaths in the UK where beekeepers have died even though such pens were issued to them. Unfortunately, it is more often the result of a failure to use the pens (plural) correctly or before the reaction has gone too far. The importance of understanding their correct use and limitations cannot be over-stated. It is vitally important that those accompanying a person carrying such pens are briefed on their use because a victim may not be in a position to administer the dose(s) themselves (N.B. the second pen is not a spare, it is a second dose).

6.4 - Summary of the Risk to Human Health

In summary, the evidence from the Spanish and Chinese is far more convincing than that from the French. The risk of dying from an Asian hornet sting is however, low. The statistical risk of a systemic reaction following a sting by a hornet, wasp or bee is also low. When it comes to the honey bee it is a risk that beekeepers have to accept as part of beekeeping as they are at increased risk of systemic reaction compared to the general population.

When it comes to the Asian hornet and beekeepers, the risk in the apiary (where the hornet can be expected to be present) is low, especially if the beekeeper is wearing their bee-suit. If the beekeeper handles the hornet they should be careful because that is inviting a sting, to which they may be unknowing vulnerable.

If the beekeeper goes looking for the hornet's nest then they are at much greater risk. Not only is there a greater chance of being stung because of the hornet's aggression and the number and location of nests, there is a greater chance of a systemic reaction because of the cross-reaction to other venoms and where this occurs, that reaction is likely to be severe. Contrary to the UK government's assurances, it therefore looks very much as though *V. velutina* venom is more dangerous than that of a bee or European hornet.

Some groups of people are clearly more vulnerable than others and should avoid contact with the Asian hornet:

- ▶ People whose health is compromised (especially with vascular and heart problems or damage to liver, kidneys etc).

- ▶ Those over 65 years old are over-represented in those having a serious reaction to the hornet's sting.

- ▶ Those who are pre-sensitised to Hymenoptera venom and have suffered previous systemic reactions (perhaps carrying auto-inject pens).

- ▶ Beekeepers, by virtue of their exposure to honey bee venom, are at increased risk compared to the general population.

Further information on insect sting allergy:
https://www.anaphylaxis.org.uk

7 - The Asian Hornet - A Consummate Predator

7.0 - Introduction

The MNHN's calendar wheel showing the hornet's seasonal cycle will be familiar to many but if the beekeeper is to understand what is happening, they need to know the hornet as well as they know their bees. Just knowing the calendar and sequence of events is not enough. The beekeeper needs to know what factors influence that cycle and associate those factors with what is happening around them and what is going to happen in the apiary.

Poster with permission of the FDGDON Ille et Vilaine - responsible for the day to day running of the Department's *Lutte collective*.

We start by looking a little closer at the Asian hornet (*V. velutina*), which is known in France as Le Frelon Asiatique; shown to the left of the European hornet (*V.crabro*). It is frequently confused with two Giant Asian Hornets, the Southern Giant hornet (*V.soror*) and the Northern Giant Hornet (*V. mandarinia*).

Method of Attack - With their huge size, powerful jaws and thick exoskeletons, it is the giant Asian hornets that make massed direct attacks on the honey bee hive and can overwhelm it in a matter of minutes; that is their main method of attack.

The much smaller Frelon Asiatique (*V. velutina*), sometimes referred to in China as a wasp, is a solitary hawker that uses its aerial agility to hawk bees in preference to the direct attack on the hive. Unlike the larger hornets it is agile enough to catch a bee in mid-air. It may make direct attacks on a hive if the hive is weakened or if the bees are clustered but it is the hawking that gives beekeepers some very difficult problems to manage.

V. velutina is an expert aerialist that can catch a bee in mid-air. High-speed film footage reveals that a hornet may even decide to overtake the bee, execute a loop-the-loop with barrel roll to come in on top of the bee in order to grab it. Photograph reproduced under licence Lothar Lenz stock.adobe.com

Our own European Hornet (*V.crabro*) is regarded as a useful part of our biodiversity, is not normally a serious problem to honey bee colonies, and ecologists are keen to avoid harm to it. Initially *V. crabro* and *V. velutina* lived alongside each other in France as somewhat uneasy bedfellows although *V. crabro* wouldn't take any nonsense from its Asian cousin.

Nowadays, and thanks to *V. velutina's* competition for its normal prey, *V. crabro* will attack *V. velutina* but there are not enough of *V. crabro* to make any difference to its numbers. Worryingly, beekeepers report that *V. crabro's* attacks on beehives are intensifying and becoming a real problem. It has been suggested that *V. crabro* is learning from its competitor but may it also be an indicator of the Asian hornet's effect on insect populations?

V. velutina's season is much longer than that of *V. crabro*, its nests are much more populous and there are far more of those nests. Furthermore, *V. velutina* nests are usually in the open whilst *V. crabro* nests are usually under cover. You are far more likely to stumble across an Asian hornet nest than a European hornet nest.

7.1 - The Asian Hornet's Seasonal Cycle

Beekeepers have to know the Asian hornet's seasonal cycle as well as they know that of their bees or they won't understand what is happening in the apiary nor will they know when it will happen and to what intensity.

Furthermore, if you understand what factors impact on that cycle you will be able to foresee what is about to happen and not get caught by surprise.

We'll start in the autumn (the purple sector – see inset graphic), this is when the hornet's nest is at its peak. Over the autumn it progressively switches from producing workers to producing sexuals, a process that is at its height in October and is usually complete by early November.z

Over a thousand sexuals are produced by an average nest (ratio 2:1 male to female). Those sexuals mate (where is not known) and the mated queens spread out to find new territory either before going into hibernation or afterwards. They hibernate for winter under leaf litter, under tree bark, under caravan awnings, in vegetable crates, lorry canvases; although researchers tell us that less than 10% make it through winter (probably fewer than 5%) in a normal year.

The queen normally dies in early November and the nest is destroyed by the winds and the weather over winter – it is only used for one season.

However, there are an increasing number of reports of nests surviving winter with live hornets in them so it should not be assumed that such a nest is 'dead'.

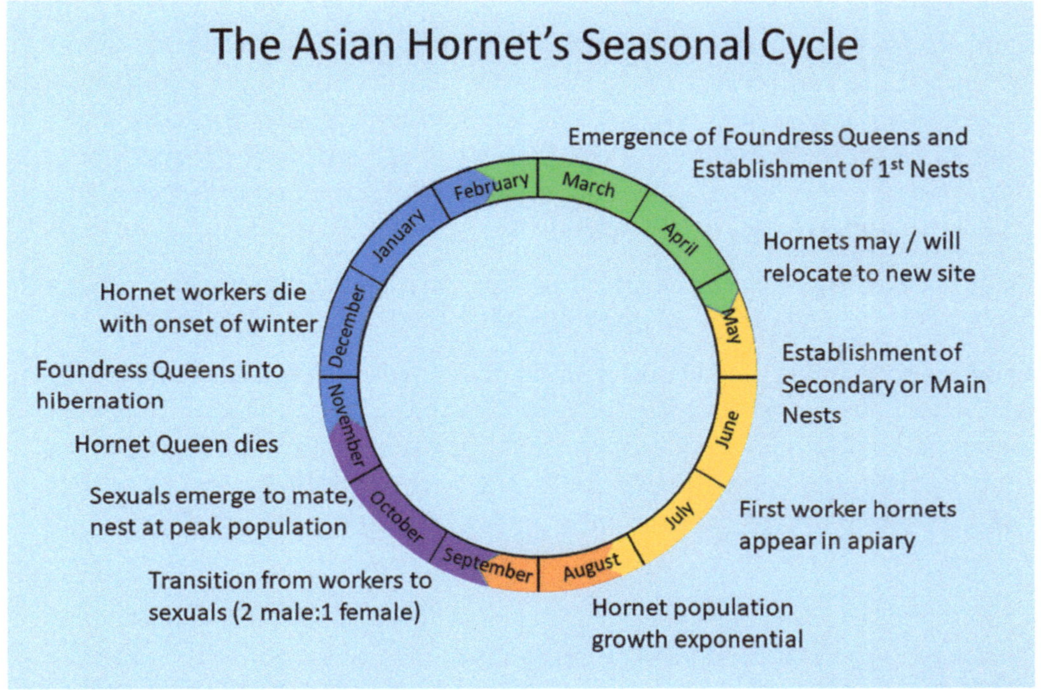

The Asian Hornet's Seasonal Cycle

The foundress queens as they are now called, emerge at different times throughout the spring (green sector). The first queens emerging when the temperature gets consistently over freezing and reaches c.12 deg. C for 4-5 consecutive days. During the first 10-12 weeks the foundress queens are on their own and vulnerable.

The first nest, the '*Nid primaire*' (also known as the 'Embryo Nest') is always under cover and hanging down by a petiole from something like a shed roof or a veranda or in an electricity meter box and if you are looking for *nids primaire* you will usually find them where last year's nests were located – so if you do find one remember where it is for next year.

The *nid primaire* grows in size and the hornets usually relocate at around the interim stage (size of a football) when the bulk of the workers move to a more favourable location to build a secondary or main nest, to be joined by the queen and other workers when it is ready. Some workers will stay behind to look after the remaining brood. If you find a small nest with no eggs or queen, they have relocated and the nest with the queen will be nearby.

Scientists estimate that, of those that survive winter, only about one in 9 emergent foundress queens actually founds a full-size colony. I am not sure what the evidence is for this estimate. It has been asserted that there is competition amongst emergent queens to take over embryo nests. This was a key argument used by the MNHN against spring trapping but two leading scientists at INRAE pointed out recently out that there is no scientific evidence for that assertion.

Over the summer the numbers of worker hornets increases and the nest grows in size until we come again to the autumn.

Predation on hives slowly builds up from July when you might see the first Asian hornet workers in your apiary but note that from September there is a marked increase in predation in the apiary and do not overlook the fact that the hornet can be very active in late October and November –those can be particularly dangerous months for our bees.

7.2 - Hornet Nest Numbers – Variation and Surges

Those who promote nest destruction often quote annual nest numbers as evidence of the success of such a strategy but many simply do not understand the fluctuation in reported nest numbers year on year, in which the weather is the primary factor. Both nest numbers and nest development are affected. Both are important factors.

The more foundress queens that survive hibernation to emerge in the spring, the more *nids-primaire* there will be, and the more *nids primaire* that survive spring to develop into secondary or main nests, the more worker hornets there will be in autumn to predate on your hives.

What causes a surge year? Take 2021, not a very good year for the hornet, nest numbers were markedly down in France but an unusually mild winter at the end of the year meant that many more hibernating queens survived to emerge in the early and warm Spring of 2022. Furthermore, there were none of the sharp frosts that often follow an early spring and so more first nests survived to develop into main nests. 2022 was a surge year, predation on hives was heavy.

7.3 - Spotting that you are in a Surge Year

Beekeepers are well-used to noting the weather's effect on their colonies and will adjust their beekeeping accordingly. They need to make the same connection to the likely development of the hornet nest. If the bees were active late into the year, perhaps the presence of brood affected your varroa treatment, perhaps there was a need to feed fondant because stores had been used up; those signs would suggest a mild winter and an increased survival rate for over-wintering foundress hornet queens.

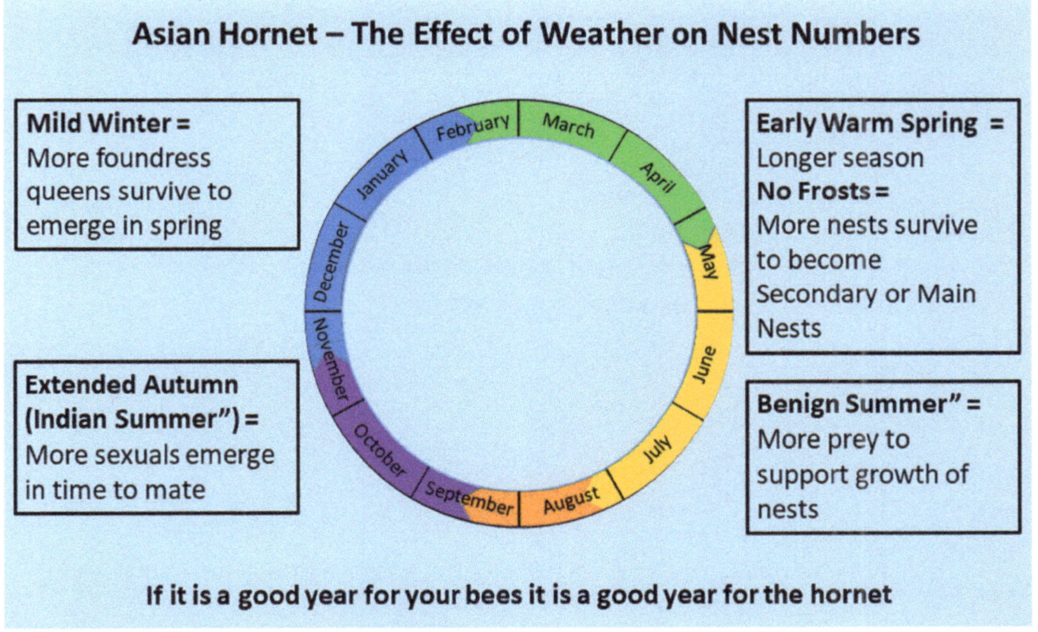

The Effect of Weather on Nest Numbers

If the bees are more active than usual in the spring, colony build-up is happening faster than normal, these are the early warning signs that suggest the hornet season may also have got off to an early start.

If the NBU are warning of unusually early nest reports and higher nest numbers for that time of year. If predation in apiaries started a month earlier than normal. Those too are your warning signs that it is going to be a surge year. You need to recognize when a surge year is happening because you will need to strengthen your defences and you will need take it into account in your management of the bees.

The triple whammy comes when summer is followed by an extended and warm autumn, you don't get any more nests but they grow bigger and more populous than usual, predation in the apiary will be heavier and last longer.

7.4 - Hornet Nest Development

It isn't just about nest numbers. How the hornet nest develops over the course of its season also plays a role.

In a Normal Year

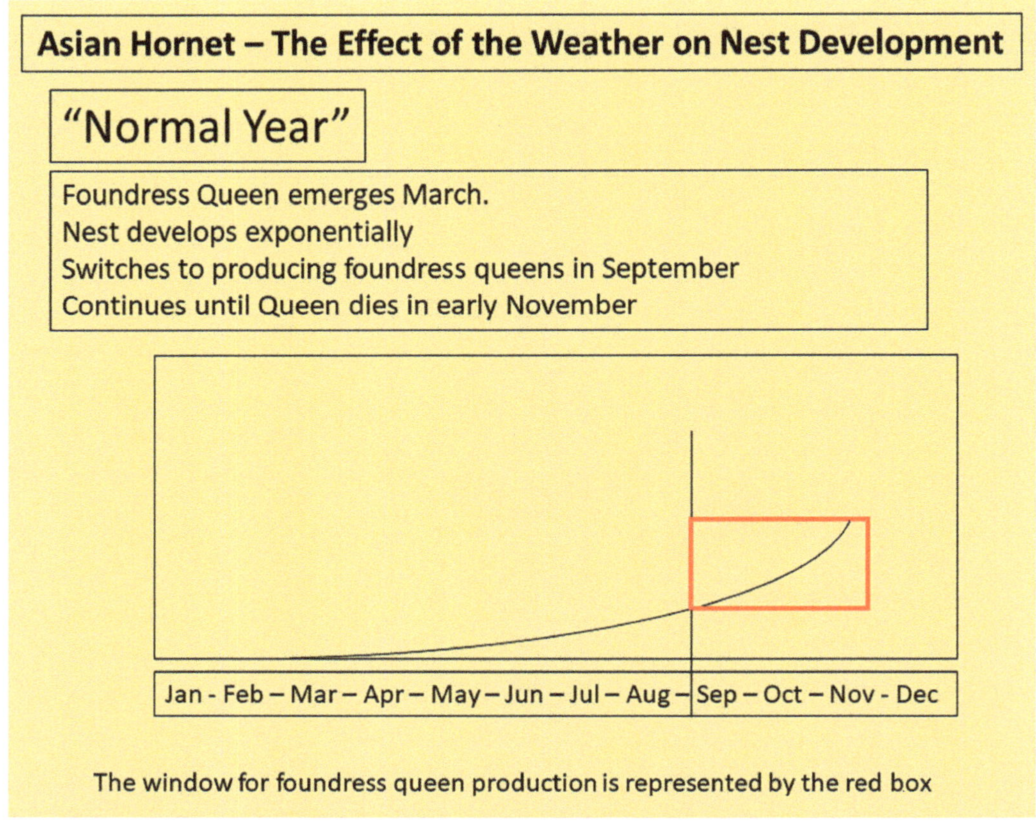

The Effect of the Weather on nest Development – Normal Year

In a normal year, the hornet's cycle starts with the emergence of the foundress queen in the spring and the nest develops over the summer. September sees the start of the production of sexuals but this stops with the death of the queen in early November.

In a Poor Weather Year

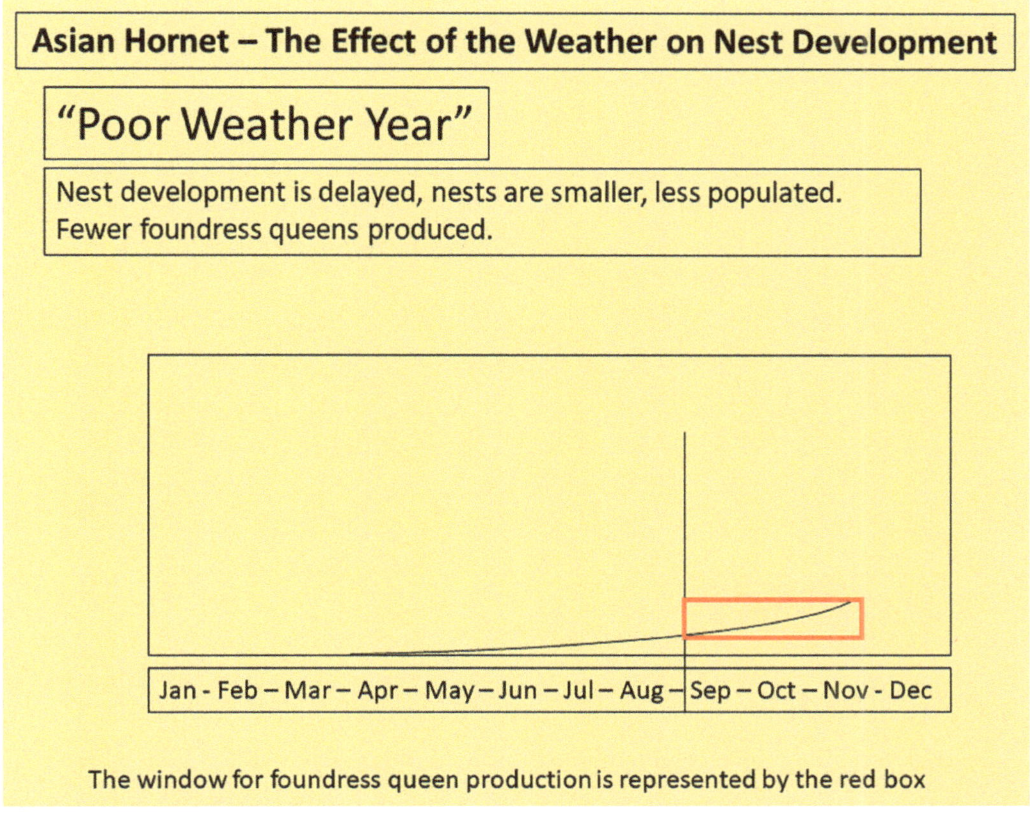

The Effect of the Weather on nest Development – Poor Weather Year

If the emergence of foundress queens is delayed, the development of the nest starts later. If the weather is poor, there is fewer prey and the growth curve becomes flattened. The nest is smaller and cannot support as much brood when the window for the production of sexuals opens, if in fact it is not delayed. If the nest cannot produce as many foundress queens, there are likely to be fewer nests the following year.

In an Early and Extended Year

In a year with an early start and benign weather for spring and summer, the nest will be fully developed and much larger and more populous by the time the window for foundress queen production opens.

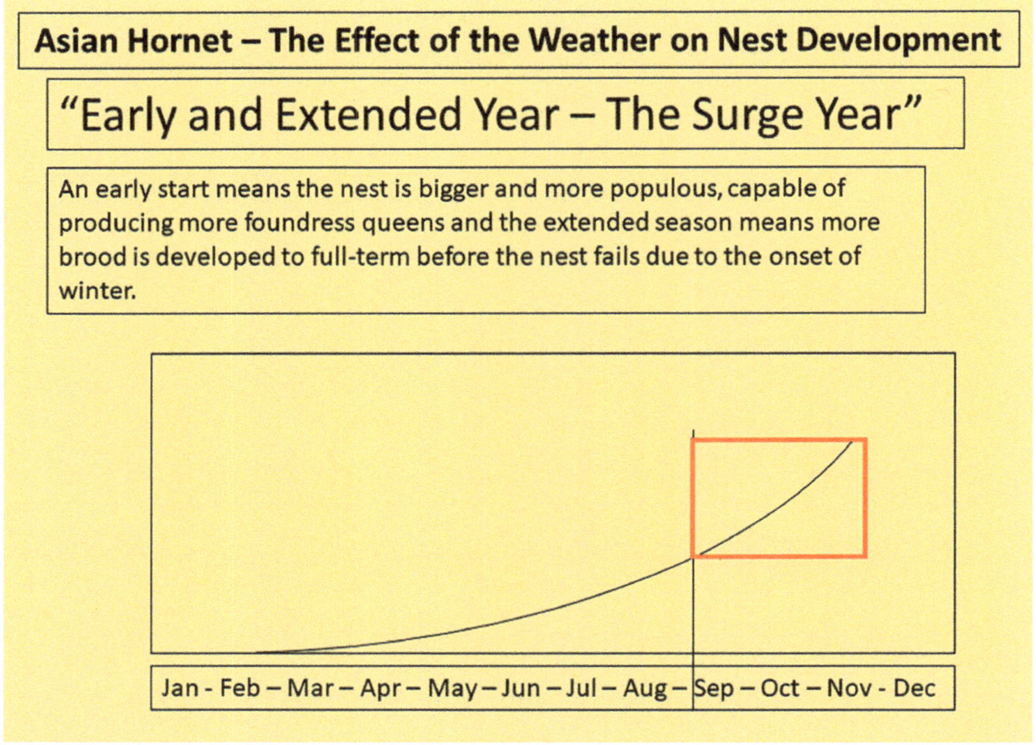

The Effect of the Weather on nest Development – Surge Year

If the autumn is warm and becomes extended (what is commonly called an 'Indian Summer'), all the sexual brood develops to emerge. That is when we get the really large nests that produce many more foundress queens to emerge, mate, and go into hibernation.

An early and extended year sets the scene for a 'super surge year' in the following year, if the weather conditions over winter and in the following spring are favourable, that is when we get a 'super surge' .

INSPIRATIONAL FRENCH BEEKEEPERS

André Lavignotte from the Pyrenees-Atlantique Department and a beekeeper for over 60 years, conceived and developed 'La *Museliere*'. The original concept came from the metal mesh screens that were used to protect cows' eyes from flies in the summer but it took a lot more work to arrive at a screen to protect beehives. La *museliere* has since been developed into many different forms, some more successful than others. André's invention has saved countless numbers of honey bee colonies.

- Photograph courtesy of Andre Lavignotte

8 - The Hornet And The Honey bee

8.0 Introduction

We are now going to look at the interaction between the hornet and the honey bee. Some of the effects of the hornet's predation are hidden – and they are the real colony killer. We need to understand what is happening with the bees.

8.1 - How the Hornet finds the Hive

What beekeepers need to pay particular attention to is that the hornet is not in the apiary by chance, it is actively searching for concentrations of its prey. It is attracted by the smell of pollen, honey and the bees' own Nasonov gland pheromone, with which the bee marks the entrance to its hive.

Once an apiary is found, the hornet learns waymarks to find it again, so a high number of hornets will be return visitors.

It is very important that you do all you can to maintain a low olfactory signature in the apiary. Extracting honey nearby or leaving hives open for longer than absolutely necessary is not a good idea. Beekeepers who spill honey around or leave out wet frames are asking for trouble. It's an extra incentive to watch bee-space like a hawk and get rid of brace comb before it gets filled with honey.

8.2 - Predation of the Honey bee

The first sign that the beekeeper will see in the apiary is Asian hornets hawking outside the entrance to a hive and, as the attacks intensify, very quickly a beard of bees will be seen outside the entrance. Remember this hornet prefers to hawk bees rather than attack the hive directly, which is why it is known as a solitary hawker.

The classic picture of a hornet stationed just outside the entrance and facing out can be misleading. The hornet likes to ambush its prey and may hide under landing boards ready to swoop in on the unwary bee.

The hornet is diurnal and returns to its nest at dusk.

V. velutina hawking a hive and staying out of reach of the bees. Note the hornet waiting to ambush bees (far right).
Photograph reproduced under licence – Rodolphe stock.adobe.com

8.3 - Effects of predation including hidden effects

Focussing now on the effect of hornet predation on the honey bee because some of the effects are hidden, and it is these that cause the real problems.

As predation builds the response of the honey bee colony is to reinforce the guard bees. More and more bees are to be seen clustered around the entrance and the colony starts to get very stressed. Although in time this response diminishes, it highlights an increased risk during the hornet's establishment phase.

What a 4-year French scientific study published in 2018 (no. 48 – see Further Reading) found was that in addition to the obvious losses caused by hawking: there was a reduction in foraging corresponding with the number of hornets in front of the hive. As a beekeeper 'rule of thumb', 4 hornets sees a 40% reduction in foraging, 6 hornets a 60% reduction and more than 12 hornets in front of a hive sees foraging paralysis. The bees stay in the hive consuming stores and getting even more stressed.

Although the study was useful in that it confirmed the beekeepers' claims that colonies were being affected by the hornet's predation, it was flawed. The study's resources did not allow adequate observation of the hives and it relied heavily on modelling. Acknowledging these limitations, the researchers said that they thought their estimate was conservative. Nowadays the general view is that total foraging paralysis comes in earlier than the report found – perhaps even as low as 4 hornets always in front of the hive. The MNHN has now defined a high level of predation as being 4 or more hornets always in front of the hive.

The first French scientific paper on the effect of the hornet's predation on the honey bee colony was based on observations taken between 2012 and 2016 of 131 bee colonies in 75 apiaries. The researchers taking another two years to publish their findings (the hornet had arrived in 2004). By focusing on what they could see outside the hive i.e., homing failure and foraging paralysis, the study overlooked the most important factor of all, what was happening in the hive.

Given that the apiaries were spread over most of the south and southwest of France, one might have expected a sizeable team to be carrying out the observations but in fact there was just one observer, 'in order to mitigate any observer effect'. A consequence was that the observation time when a hive was visited was very short, just 17 minutes.

Just one video camera was used on just one hive but with a narrow field of view and so it was unable to track predation other than that immediately in front of the hive.

The 'BEEHAVE' computer model used to predict survivability did not take into account the need for the production of the winter bees essential for colony survival. Hives were not opened or other monitoring devices used.

The scope of the study and the resources available for research on this scale had been totally inadequate. The study was better than nothing but not by much.

Although homing failure and foraging paralysis were identified as effects of the hornet's hawking, with a consequent reduction in the colony's survival probability, there was one other very important factor that the researchers conducting the French study said that they had not been able to assess and that was the effect of the hornet's predation on brood production.

It took beekeepers to point out that they were suffering increased queen loss, autumn swarming / absconding. Beekeepers know that at times of reduced food income there is a reduction in brood production, if in fact the colony does not stop the queen from laying any eggs. Not all understood that another consequence of this was a failure of the bees to make their winter bees.

The graphic (inset below) from Randy Oliver explains this very well. Reds and yellows are summer bees, the blues and purples are winter bees, the bees with longer lifespans with hypertrophied pharyngeal glands and enlarged fat bodies, that are essential for the survival of the colony over winter, The dotted line shows the brood production. Beekeepers will recognise the peak for spring swarming and another to get the colony in peak condition for the main honey flow. Then there's a sharp drop that, especially in my Carniolan colonies, reflects the drop in food income. This is a time when beekeepers often report more defensive colonies and I believe that defensiveness is a reflection of the stress that the colonies feel on finding that their stores for winter have disappeared with the beekeeper.

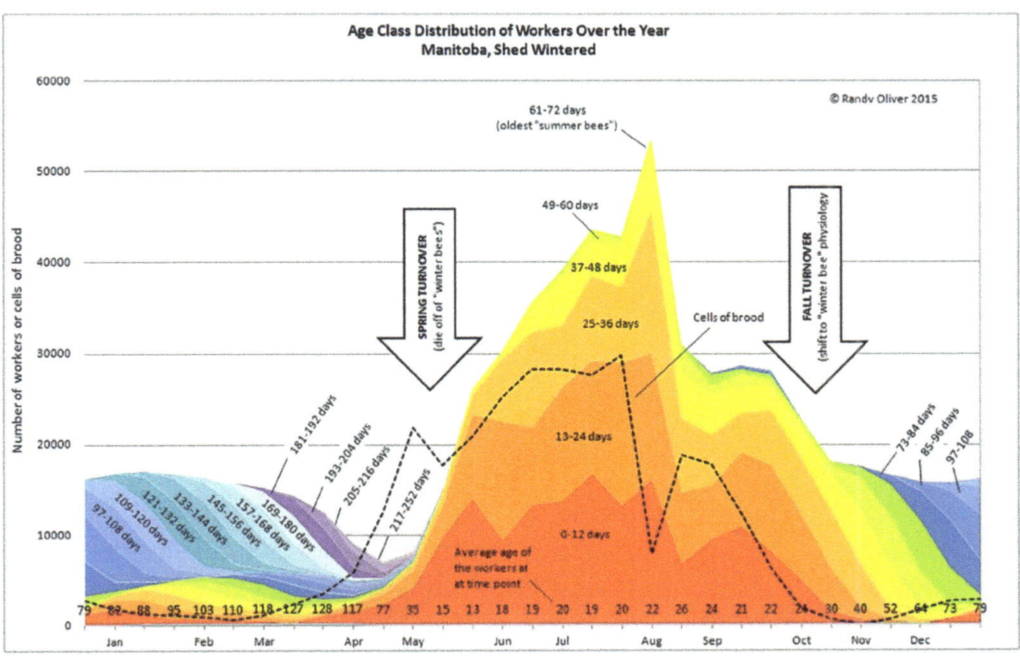

Graphic by kind permission of Randy Oliver – Scientific Beekeeping.com © Randy Oliver 2015

Mattila, et al (no. 50 – see Further Reading) found that the bulk of a colony's true winter bees are produced in that last peak, in one short period in the autumn, and we know from the hornet's seasonal cycle that this coincides with the peak Asian hornet predation period. That is why it is in winter that colony losses from the hornet are highest.

To summarise the Asian hornet's predation:

- ▸ There are two methods of attack, hawking and the possibility of direct attack in the autumn

- ▸ Predation starts from a low level in July but with a significant increase in late August/September and peak predation in the apiary in October, and it can continue into November. This diverts foraging bees to guard bee roles and put the colony under great stress.

- ▸ Foraging is reduced as more returning foragers issue 'stop signals' until the hive is in a state of foraging paralysis (no. 70 in Further Reading).

- ▸ The loss of bees hawked by the hornet weakens the colony but a healthy colony can usually survive this.

- ▸ Starvation, but the beekeeper can do something about that and hopefully didn't take all the honey off at the end of July and leave them with no stores. Sadly, it is obvious that some beekeepers do not act to make up the lack of stores. Not only is this the case in France, the regular e-mails from our own NBU in August is evidence that UK beekeepers are just as culpable.

- ▸ Don't under-estimate the effect of stress on the bees which can lead to colony collapse, bees to abscond or kill the queen.

A more recent Chinese paper published in 2023 examined the effects of Asian hornet predation on *Apis mellifera ligustica* colonies, comparing them to the effects on *Apis cerana*. It highlights differences in reaction to the hornet's predation between *Apis cerana* and *Apis mellifera* and the effects on foraging and egg-laying by the queen leading to colony collapse. If a beekeeper has any doubts about the hidden effects of the hornet's predation, then this is an excellent paper to read and it can be accessed through the ResearchGate website (no.70 in Further Reading).

It is very important that the beekeeper knows when their colonies are at greatest risk and understands the hidden effects of predation.

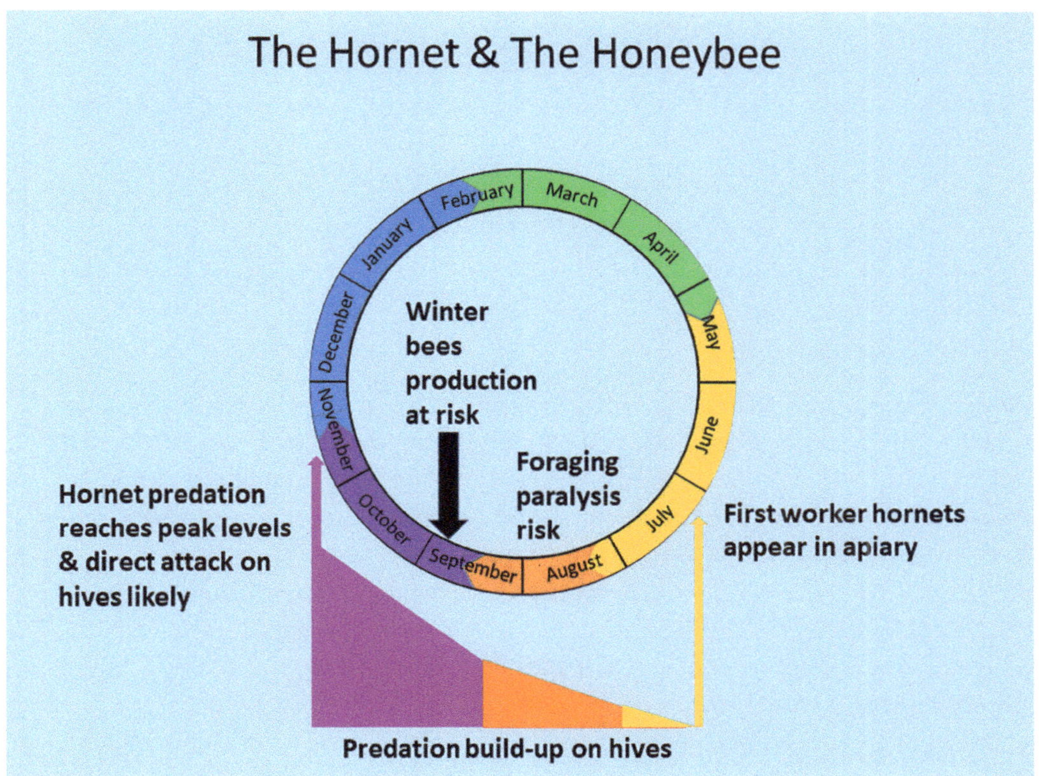

The Hornet and the Honey bee Calendar

Some losses due to the hornet's predation are inevitable, however the constant comparison of *Apis mellifera* with *Apis cerana's* Mexican waves and the balling of hornets, underestimates the honey bee's capacity to learn.

'The attacks on the hives were visibly more marked this year than last year. However, an interesting observation should be noted, the bees tend to defend themselves better against attacks by hornets' - FDGDON 56 – 2018

9 - Asian Hornet – Honey Bee Colony Mortality, Good Beekeeping, and Healthy Colonies

9.0 – Introduction

We often hear accounts of colony losses that are attributed to the Asian hornet. It is essential that we get the losses into perspective or you can fall into the trap of thinking that the hornet cannot be managed. As we saw in section 4 (Scientists and Solutions) there have been attempts to quantify those losses and put a monetary value on them but the fact is that there are too many factors involved in honey bee colony mortality losses and the situation is further complicated by the fact that a beekeeper's losses are also affected by an area's favourability to the hornet and the weather.

In some respects, it is best to think of the Asian hornet as 'the straw that breaks the camel's back'.

'The bee mortality phenomenon is multifactorial, which makes its study and prevention all the more difficult'

Gilles Salvat
Deputy CEO in charge of animal health and welfare – ANSES July 2021

9.1 - Honey bee Colony Losses Due to the Hornet

Both ANSES (French Agency for Food, Environmental and Occupational Health & Safety) and INRAE have warned that it is not possible to attribute losses directly to the hornet due to a lack of data and because colony losses are multifactorial.

In a 2021 interview with UNAF's 'Bees & Flowers' magazine, Denis Thiery, the Director of Research at the INRAE Bordeaux campus, estimated that the hornet was responsible for 30% of colony losses in affected areas (i.e., 3 out of every 10 hives that are lost, are lost due to the hornet), which is not the same as saying that the hornet is causing a 30% loss of colonies and he qualified his statement by saying that data was sorely lacking.

Losses attributed to the Asian hornet (shaded in red) are estimated to be 30% of colony losses (shaded in blue). 2018-2023 overall national winter colony losses were circa 26% (ESA Survey).

A measure that is commonly used to assess honey bee colony mortality is winter losses.

The most authoritative statement of French honey bee colony winter losses is the ESA survey that has been published each year since 2018. Covering the losses over the winter of 2023/4, the national average of dead colonies was 18 ½% and overall winter losses (i.e., including colonies that were dead, too weak or damaged to be recoverable) had risen to 31%. These are losses due to all causes not just *V. velutina*.

31% winter losses compared to c.26% for each of the previous 6 years was not as bad as had been feared. But for the first time the Asian hornet was the beekeeper-cited top cause of loss (c.21% of total losses) with 'colonies too weak to overwinter' coming in a close second. However, the category 'too weak to overwinter' likely includes an element of weakening due to the hornet. Nevertheless, these losses are multifactorial.

Winter losses do not include colony losses due to direct attack but those are far less and, in many cases, could have been mitigated or even avoided completely if the beekeeper had deployed even some of the more basic defences.

There is the argument that beekeepers overreact to a perceived risk to their colonies, implementing measures such as trapping when they are not necessary, thus putting biodiversity at risk. An argument that was put forward by four leading scientists, who pointed to the actions of stakeholders (beekeepers) trying to protect their colonies in advance of science-based evidence and recommendations.

Given that the hornet arrived in France in 2004 and that French research has taken years to conduct field studies of doubtful quality and even more years to publish the results, one might reasonably demand that the scientists themselves admit some responsibility for the situation that they are complaining about (no. 23 - see Further Reading: 'Science communication is needed to inform risk perception and action of stakeholders').

Looking ahead to the hornet's establishment in the United Kingdom, we must ensure that we beekeepers do not over-react but at the same time, the scientists must understand that we cannot afford to wait for the science to catch up.

9.2 - How the Threat from the Asian Hornet Fits in With Other Factors Affecting Honey Bee Colony Mortality

Let's look at the factors behind colony mortality in France more closely. They come from a report by a committee set up in 2017 by the French Senate to investigate honey bee colony losses (no. 57 – see Further Reading). The committee's investigation was held in the context of other reports and the findings mirror a much earlier report by the AFSSA (Agence Francais de Securite Sanitaire des Ailments) in 2009 on the 'Weakening, Collapse and Mortality of Bee Colonies' (no. 56 – see Further Reading).

Pathogens, pesticides and varroa were identified as the top three concerns; although climate change has now joined the top three and has had a drastic effect on the summer nectar flow in many southern and SW areas.

Union Nationale de L'Apiculture Francaise

'for a good fifteen years the blooms are getting earlier and faster. From the month of July, the season is over whereas previously it stretched over several weeks in summer. '

Deprived of nectar at the end of summer, the colonies suffer. 'Many beekeepers wonder if their hives will survive the winter in good condition. In addition, the hornet weakens the colonies, making it doubtful they will survive over-winter. '

Before the 2022 surge in nest numbers, the hornet was a second-tier concern alongside other factors such as the effect of poor apicultural practices and the loss of forage due to increasingly intensive agriculture.

Poor apicultural practice was a 'catch-all factor' that covered anything from beekeepers tolerating weak colonies to a need for better knowledge. Unfortunately, it does also include a failure to act and a report for FNOSAD in 2023 saw almost half of GDSA's calling for better knowledge of the hornet and defences, the lack of implementation of which was exposed in the 2022 surge.

At the Senate hearing, the National Institute for Agricultural Research was keen to emphasise that all of the factors identified were cumulative and furthermore, one factor plus another did not equal a doubling in effect. One factor on top of another was more akin to an increase by a factor of ten! The problem in trying to assign colony loss to the Asian hornet is that it is often 'the straw that breaks the camel's back' coming last and on top of the other factors.

Even after a second consecutive hornet surge year in 2023, the majority of winter colony losses were down to factors other than the Asian hornet. Factors that even if they are not within the control of the beekeeper, are factors that the beekeeper can act to mitigate.

The graphic below is extracted from the 'Platforme ESA' winter loss survey data of 2022/3; the winter that followed the surge year of 2022, with its high hornet predation on apiaries. Some 60,000 beekeepers were contacted and invited to participate in the survey. 18,000 beekeepers took part and it is notable that a third declared that they had had no losses over winter. Beekeepers who had lost colonies were asked to identify what they thought was the principal cause of loss and some 9,000 beekeepers in the category for apiaries of under 50 hives answered that question (losses for the 50> hives categories were slightly different).

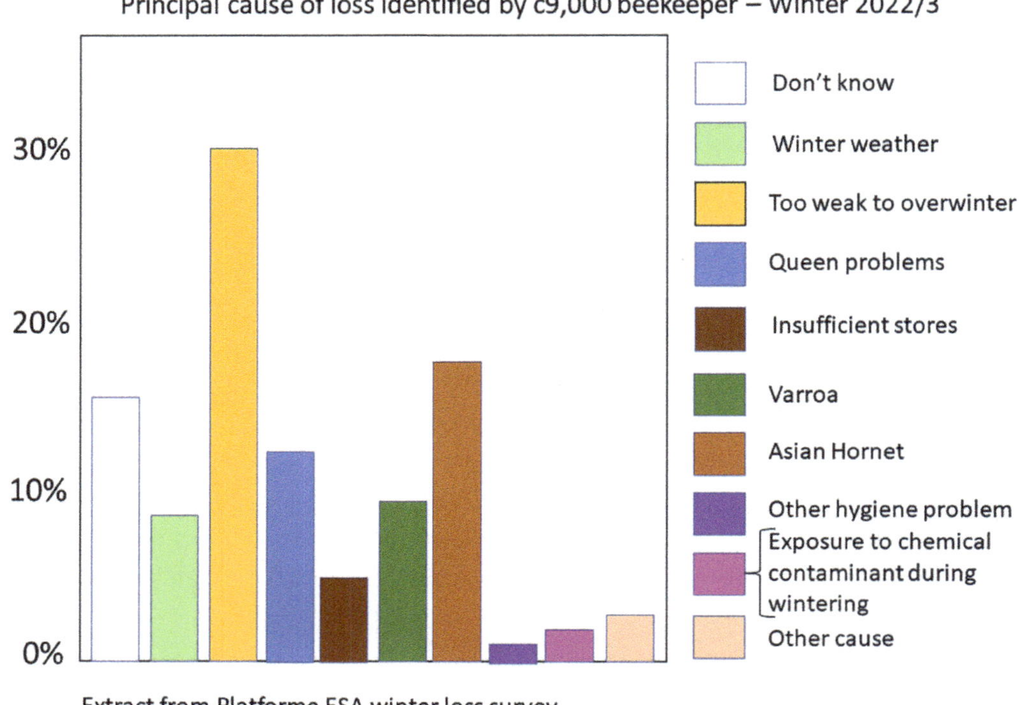

Principal Causes of Colony Loss Winter 2022/3 (extracted from ESA Survey)

18% of winter colony loss was attributed to the Asian hornet and 10% to varroa (this was much higher in apiaries of 50> hives where losses to varroa equalled losses to the hornet). The top cause of loss at 30% was attributed to colonies 'too weak to overwinter'. Queen problems accounted for 12%.

Although some departments in infested regions have above the national-average winter losses (e.g., Cote d'Armor), others such as Morbihan in Brittany have below national-average losses. There is no direct correlation with the Asian hornet and many above-average loss departments suffer from other factors such as climate change and pesticides.

When a colony is lost because it is too weak to overwinter and the beekeeper explains that the Asian hornet is to blame; the question arises, was the loss down to the Asian hornet or was it down to the fact that the beekeeper had taken all the honey off at the end of spring, then the expected summer flow had not materialised due to climate change but the beekeeper had not intervened to feed their bees? The colony went into the Asian hornet's predation period with low stores and thoroughly stressed. A prime candidate for overwintering failure.

The critical question for the beekeeper is therefore not just colony losses attributable to the hornet but also losses due to other causes that the beekeeper can influence and mitigate. We will look at that in the next section under 'management of the bees'.

9.3 - Management of the Bees

We now come to one of the most important elements of defence in the apiary – the beekeeper. It is not the role of this book to give general beekeeping advice, except to point out where beekeeping directly interfaces with the Asian hornet.

In the overall summary section (Section 15) I have given a number of scenarios, i.e., hornet reported in your region, in your area and when it is in your apiary predating your bees. I have also suggested defences appropriate for whether predation is low/medium or heavy.

Some measures need to be thought through in advance but there is no point making life difficult for yourself and your bees until the situation warrants it. I wouldn't rush to close down to a single bee entrance just because there are some hornets in the apiary, that's what the *museliere* is there for (section 12.4).

If the hornet is not in your area and you haven't seen any in the apiary, why not open hives during the day, just don't set up a situation where they find it because you are careless. If there are hornets in your area you should have planned your health inspections to take place outside the predation window. If you are doing swarm prevention in the latter half of July, something has gone wrong!

9.3.1 - Varroa Treatment

Varroa weakens bees making them easy prey for the hornet and it spreads pathogens especially when bees drift to other hives – so the beekeeper needs to do something about varroa and also drifting (see section 12.3 - Defences in the Apiary – Apiary Environment for more details on the latter). There are many different varroa treatments and it is well worth paying special attention to which one to use during the period when the hornet is predating the hives or the beekeeper could make matters worse for the bees.

As a general rule, the faster the treatment (formic acid pads for example) the more aggressive it is. You may recall how NOD, the manufacturer of MAQ's (now Formic Pro), advised beekeepers to do their regular inspection, let the bees 'put their affairs in order', then go back into the hive on another occasion, slip in the MAQ pads and close up the hive with minimal disturbance. This was all to do with minimising stress on the bees (sadly, I doubt many beekeepers went online to get the manufacturer's full datasheet).

A book on managing the hornet recommended the application of Formic Pro during July, following the removal of the honey crop, but leaving a minimum of one full super on the hive. It also recommended reducing the entrance to just one bee space during the day. When it comes to the hornet we need joined-up thinking. Let's examine that recommendation in more detail.

Leaving a super on the hive to avoid a reduction in brood production (following a sudden drop in food income and colony reserves) is, I believe, sound advice (see Stores & Feeding below) but what about the use of Formic Pro at this time, is that fully thought-through? The beekeeper needs to take into account:

1. Authorization for the use of this product in the UK requires removal of supers with honey during treatment with Formic Pro and if the super is left on during the treatment, it must then be removed and not enter the human food chain. Mark the frames to ensure they don't get mixed up with others if you leave a super on the hive.
2. The manufacturer's requirement is for full-width entrance ventilation. July is usually early in the predation window and the hornet's presence outside the hive entrance is still building.

3. The treatment has a temperature range during application of 10 – 29.5°C. Hot temperatures, especially >33°C., during the first 3 days may lead to excessive bee, brood and queen loss.
4. The treatment may trigger supercedure of fragile queens regardless of age

The 80% Colony Loss - A Myth is born

A scientific paper included the following statement:

'... beekeepers estimate that that they have lost between 5 and 80% of honey bee colonies (average 30%) where V. velutina has established...'

In attempting to summarise a French conference paper, *Monceau K, Thiery D, Vespa Velutina - Current Situation and Perspectives 2016* (no.58 in Further Reading), the author of that statement was misquoting the source, which actually read:

'In one report, a beekeeper claimed to have lost up to 80 % of his hives due to V. velutina predation (Cazenave 2013). In Gironde (southwest France), 30 % of hives were weakened and/or destroyed by V. velutina in 2010 (http://www. unafapiculture.info). In Dordogne, an epidemiological study (questionnaires sent to beekeepers) of 1,979 hives in 2009 and of 1,991 hives in 2010 indicated that ca. 5% of the hives were destroyed by ......'

Just one beekeeper reporting an 80% loss and 30% 'weakened and/or destroyed', which was not an average 30% and is not the same as lost.

That scientific paper has been taken by a number other researchers as an authoritative statement of French honey bee colony losses due to the hornet and so the myth was born of up to 80% losses and a 30% average loss and continues to this day...

Fast treatments of the vapour type carry a greater risk of brood loss (not the idea at all during hornet predation) and even queen loss, which would be a disaster. I suggest it is not a good idea to close down an entrance when a fast treatment is on the hive but neither is a full width hive opening or offset supers when the hornet is hawking around the hive, even if you have a *museliere* fitted. To my mind the biggest argument against such a treatment is the additional stress that it places on the colony. Remember INRAE's admonition that one factor coming on top of another is not just 1+1=2 but 1+1=10 in terms of the effect on the colony.

Even the slower (i.e., longer) vapour treatments such as ApiGuard (Thymol crystals) need care because if it is too warm that can cause the same problem as the Formic Acid treatment. In addition, and although Apiguard doesn't now say that you cannot feed with the treatment on the hive, the manufacturer's advice on this becomes rather blurred because it always used to say don't feed because it may interfere with the efficacy of the treatment, and feeding is what you may need to be doing if predation becomes severe.

You may have an objection to so called 'hard' treatments, as synthetic chemical treatments such as ApiVar (Amitraz) are sometimes referred to, but they are less stressful for the bees and you can feed at the same time.

9.3.2 - Apiary Hygiene

The beekeeper can spread pathogens through poor apiary hygiene and weakened bees are easy prey for the hornet. It is not in the scope of this book to go into apiary hygiene in detail but you should expect periods of confinement of the bees to the hive under heavy hornet predation. In the hygiene context that goes to clean comb not harbouring viruses and extra space to ease stress and reduce virus transmission due to congestion, etc..

Wax left in the apiary attracts the hornet (see olfactory attractants – section 8.1 - Hornet and the Honey bee – How the hornet finds the hive).

Not spilling honey or syrup in the apiary is especially important when the hornet is in your area as it too will attract them.

9.3.3 - Unite Weak Colonies / Re-Queen early / Keep Reserve NUCs

It is clear that weak colonies do not fare well under predation by the Asian hornet. INRAE advise beekeepers to ensure that weak colonies are united before hornet predation starts.

The hornet's predation period is not the time to have a colony decide to supercede. Virgin queens flying out to mate are unlikely to make it back to the hive. The beekeeper may want to re-queen early or to make some nuclei to keep in reserve; so that they can quickly re-queen a colony that has gone queenless and using a nuclei is a safer option. It is better to put nuclei next to or between strong colonies so they don't get singled out by the hornets.

9.3.4 - Stores and Feeding.

The honey bee colony stores far more than it requires for its immediate needs in order to build reserves. Removal of reserves prompts the colony to work hard to replace them but if that coincides with a period of dearth of nectar or an inability to forage, the colony becomes very stressed. Stress in the predation period is to be avoided at all costs.

The hornet's predation typically starts to build up to problem levels from mid -August onwards. This often coincides with the time that many beekeepers have taken off their summer honey and are putting on a varroa treatment that requires the removal of supers. The bees are left with inadequate stores and hornet predation, a very bad combination.

For years I have adopted the practice of always leaving a super on the hive and nadiring it in the autumn. It stays on during varroa treatment and the frames are marked with an 'X' to show that they cannot then enter the human food chain. That super is the first to go back on top of the brood box in spring, by which time it is always completely clean of honey.

E.H.Thorne Robson Feeder (left)   Homemade dual feeder (right)

Whether you do this or not, you must be ready to feed the bees both syrup and pollen if you are to ensure that the bees produce their winter bees, essential for the colony's survival. This may require the use of a dual feeder so that both syrup and pollen cake can be fed at the same time. E. H. Thorne sell such a feeder (Robson feeder - £50 in 2024) or you could modify a spare cover board to take an Ashforth feeder using an eke.

Do not forget to have a water supply close to the hives (or between them) or you could modify an OMF (Open Mesh Floor) to take a water feeder. Get the bees used to the water source so that they don't feel tempted to use the water bath under a *harp electrique*.

Hive fitted with entrance restrictor and water feeder. Photograph by author

9.3.5 - Do Not Stress the Bees

The one word used by all French beekeepers to describe bees under hornet predation is 'stressed'.

It is because of the stress factor that the French national bee health association urged beekeepers to use solid floors and to be very cautious about opening the hive during the predation period. However, we use OMF's for a reason and so an alternative suggested by the AAVO (Friends of the Bees of the Val-D'Oise) is to screen off the underneath of the OMF by placing a super fitted with a second OMF-screen (or a mesh) under it (see section 12.3 - Defences in the Apiary – Apiary Environment for an alternative to that idea).
Don't forget to screen off the back of the OMF.

Beekeepers need to get intensive hive manipulations out of the way before the hornet's predation in the apiary builds up to become a problem. Typically, that's by the beginning of August, but in an early hornet season it can be the beginning of July – you can see how that can make life a lot more difficult for the beekeeper.

The hornet needs to get going in the morning and so predation tends to peak around midday. In summer, especially if the weather is hot, the hornet will get going much earlier and avoid the midday heat, and so predation falls into two periods, one in the morning and one in the afternoon/early evening.

In the period of predation, hive manipulations need to be minimised and if required done very early in the morning or just before dusk.

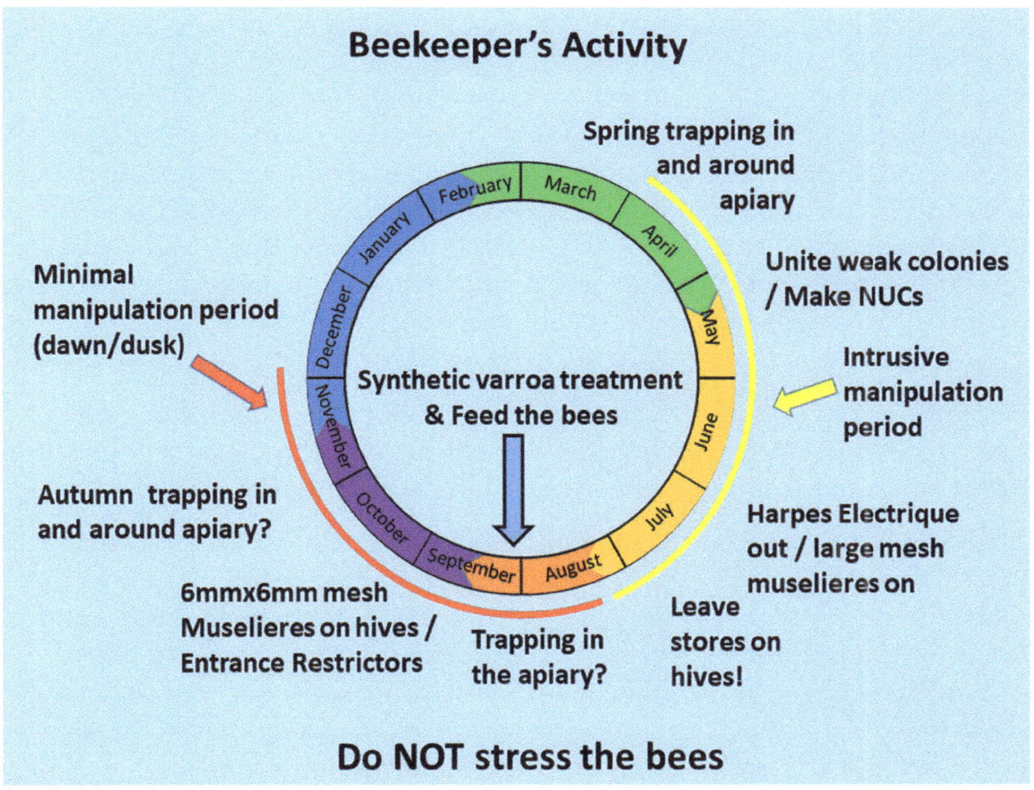

Beekeeper's Activity Calendar

10 – The Spring Trapping Debate

10.0 - Introduction

We must now consider what has proved to be one of the most controversial aspects of the fight against the Asian hornet and we should start with the case made against it. Although the spring trapping of foundress queens has been conflated with nest destruction in many departments of France, I looked at it in the section on nest destruction purely from the viewpoint of does it work and should it be part of a wide area integrated control strategy. Many beekeepers have very strong views on the trapping of insects and so in this section I look at the debate that took place in France, as a precursor to its inclusion in beekeeper defences in the vicinity of the apiary.

There is no such thing as a 100% selective trap and even if a trap is designed to be as selective as possible, with escape slots to allow by-catch to escape, there are entomologists that argue that even a short time spent in a trap will damage the insects necessary for our biodiversity.

The MNHN has always argued vociferously against any form of trapping, fearful that the public's reaction to the hornet would result in widespread indiscriminate trapping that would destroy insect biodiversity. Early on in the invasion the MNHN posted a photograph of the catch in a non-selective trap. For only three Asian hornets there were over a thousand other insects caught. That 2009 photograph was used to beat beekeepers over the head whenever spring trapping came up and it is to be found on the MNHN website even to this day.

An anomaly becomes obvious when one studies the catch in that photograph. Thirteen European hornets against three Asian hornets, at least one of which looks small enough to be a worker, suggests that the trap was out far too late in the season for Asian hornet queens. By the time that trap was emptied, the Asian hornet queens would have been confined to their nests in the role of egg-layer. Furthermore over 94% of the catch were flies that would be readily attracted to a drowning trap which clearly had no escape holes. It was not a selective trap yet the beekeepers have always been at the forefront of the move to selective trapping.

A 4-year study was initiated in a number of departments to assess the effectiveness of spring trapping, some of the departments implementing wide-area spring trapping, others not.

As part of the study the traps were regularly emptied, the contents identified and reports submitted showing the number of insects caught; broken down into species (Asian hornets, European hornets, wasps, moths, flies etc). As can be seen by the figure given for Asian hornet queens caught in the Morbihan *Lutte collective*'s spring trapping (see inset below), the number of foundress queens caught was enormous.

It was the scientists' contention that only 1 in 9 emergent foundress queens actually succeeds in founding a main nest but the numbers caught vastly outweighed nest numbers subsequently reported. Either far more embryo/primary nests fail or there are far more unreported main nests than has been realised, or both. The MNHN argued that the explanation was competition between queens but without any scientific evidence to support that contention.

10.1 - The Morbihan Wide-Area Spring Trapping campaign

In Morbihan the spring trapping campaign had actually started in 2015, a year earlier than the official study. In the spring of 2016 (the first year of the ITSAP study) Morbihan's campaign caught almost 50,000 foundress queens yet they still had 5000 reported nests suggesting that they had only caught half the foundress queens (5,000 nests x 9 = 45,000 emerged queens; in addition to the 50,000 queens caught in spring trapping?). Even so, that was a real achievement compared to number of reported nests in adjacent departments of Finistere (7,400 nests) and Ille-et-Vilaine (6,200 nests). It was an apparent endorsement of the benefit of spring trapping. Yet there must have been a large number of unreported nests for in the spring of 2017 Morbihan caught 68,000 foundress queens. Which again suggests that even a well-organised campaign was destroying less than half the number of actual nests.

Year	Secondary nests	Change	Notes
2020	3,570 secondary nests	(+25%)	(No trapping figure - 433 Primary nests destroyed) — France Bleu Report 30 March 2021
2019	2,866 secondary nests	(-31%)	(Spring 2019 – 23,383 queens + 500 Primary Nests) Urban density 5 -13 nests/sq.km / Rural density 0.3 – 0.4 nests/sq.km ?
2018	4,178 secondary nests	(+40%)	(Spring 2018 – 41,633 queens + 590 Primary Nests)
2017	3,089 secondary nests	(-40%)	(Spring 2017 – 68,000 queens + 700 Primary Nests)
2016	5,000 secondary nests	(+50%)	(Spring 2016 – 49,072 queens)
2015	3,500 secondary nests	(+300%)	(Spring 2015 – 24,612 queens)
2014	1,150 secondary nests	(+500%)	
2013	235 secondary nests	(+400%)	
2012	63 secondary nests		
2011	Hornet reported in Dep't		

Spring trapping for a limited period started in Morbihan in 2015. The number of queens caught suggest either a higher attrition rate in the *nid primaire* stage or (more likely) that there are far more nests in the countryside that has been realised.

The summer of 2017 saw heatwaves, to which were attributed the low number of reported nests that year. Again, there had to have been unreported nests for the next year, 2018, they caught 41,000 queens. We should however note that FDGDON 56 reported difficulty in getting the referents' reports and there were undoubtedly more queens caught than recorded.

2018 saw a surge in nests in France but Morbihan reported only 4,178 nests, again a remarkable achievement compared to adjacent departments. Post 2019 reports were more difficult to access and so 2020 is a Press report.

10.2 - Ille-et-Vilaine, Brittany –
The conclusion after many years

Ille-et-Vilaine is the eastern-most department of the four Brittany departments. In 2023 it destroyed 8,722 nests as part of its department-wide nest destruction scheme. The scheme had started in 2014 and for a number of years had included wide-area spring trapping of foundress queens.

Ille-et-Vilaine abandoned spring trapping, concluding that:

▸ Monitoring of trapping became chaotic from the second year.

▸ No control was possible in terms of intensity and practices.

▸ The argument that one hornet queen caught equalled one less nest could not be proved. They caught far more foundress queens than there were ever reported nests.

▸ The lack of selectivity of traps gave rise to concerns for the general biodiversity.

▸ They were unable to correlate the results with nest numbers as the variations in nest numbers year on year could always be explained by natural and meteorological conditions.

Sadly, their concerns over non-selective trapping were borne out by a nation-wide FNOSAD survey in 2023. None of which was to say that spring trapping didn't work, merely that they couldn't control it. FDGDON 35 were however adamant that spring trapping in the vicinity of apiaries by beekeepers (and I quote) *'must be maintained as it reduces pressure around the hives'*.

Fortunately, we don't have to decide whether it was the trapping, the weather, the number of unreported nests, competition between queens, or a combination of all of these factors. At the end of the study the FDGDON 56 (Morbihan) concluded that spring trapping and a campaign to find the embryo nests (*Nids primaire*) had made a significant difference.

10.3 - The ITSAP Analysis

An analysis by ITSAP of the data from the departments in the trial (published as a Technical Note) agreed with the FDGDON conclusion but with caveats about the use of selective traps and trapping only for a limited period (ITSAP Institut de L'Abeille 'Note Technique - Lutte contre le frelon asiatique (Vespa velutina) Piégeage des fondatrices au printemps' –

5 February 2021). The MNHN had little alternative but to agree the finding, with grave reservations about such trapping. However, it wasn't going to give up its fight against spring trapping.

As already detailed in Section 4.4 (Beekeepers under the Spotlight) the MNHN had pre-empted the ITSAP announcement with their paper 'Science communication is needed to inform risk perception and action of stakeholders' (no.23 - see Further Reading) accusing the beekeepers of acting without the benefit of scientific advice and implementing measures to protect their colonies with a distorted view of the actual risk.

It was to be followed in 2021 with the release of another a study that had been conducted years earlier (2008 – 2010) by MNHN and CNRS researchers, described in the paper 'Not just honey bees: predatory habits of Vespa velutina (Hymenoptera: Vespidae) in France' (no. 61 – see Further Reading), which had established that a single hornet nest consumed around 11kg of other insects (circa 100,000 insects) in a single season.

> Well-organised trapping of foundresses in the spring and careful observation of suitable locations for the establishment of primary nests considerably reduce subsequent observation of secondary nests.
>
> FDGDON 56 (Morbihan) conclusion at end of its 5-year campaign)
>
> ITSAP analysis – 'Trapping *Vespa velutina* foundresses in spring reduces the number of Asian hornet nests provided that it is repeated over several successive springs. Initial results show that with continuous trapping the effect on the number of nests is reinforced (for example, trapping repeated without interruption for 4 years is twice as effective as when it is carried out only over 3 years)'
>
> Technical Note issued 5 February 2021

Put the two papers together and what the MNHN was saying was that the beekeepers were acting irresponsibly and that it wasn't just honey bee colonies that were suffering.

The ITSAP Technical Note was a second 'victory' for UNAF, who were strongly in favour of spring trapping but not everyone accepted the ITSAP finding as 'scientific'. In her Asian Hornet Handbook (2nd edition), Dr Sarah Bunker states that ITSAP's confirmation was 'insufficient' evidence to prove that spring trapping works i.e., the preliminary findings from the study did not meet the scientific standard of proof. Furthermore, the results have not been published in a scientific journal, and therefore there is still no clear proof that spring trapping works. It's a peculiarly academic view.

The study was supervised by ITSAP, who brought in a statistician to review the results from the departments; but Dr. Bunker's contention raises an interesting question; why was there no follow-up paper? I would speculate that as the findings went very much against the views of the powerful cohort that dominated that area of science in France, it would be a brave scientist that put their name to a paper and furthermore, who would do a peer review, the same cohort?

Would ITSAP, founded in 2010 but by 2017 struggling for its very survival and in receipt of emergency state funding, want to be associated with a paper that went against the MNHN, the government's advisor on the Asian hornet? But that still doesn't mean that the study and subsequent review were meaningless, the sheer number of queens caught each year spoke for itself.

The FDGDON that reached the conclusion on spring trapping in Morbihan, are a pest control organisation working on a number of invasive species programmes. Meticulous in their reporting, neither beekeepers nor scientists, they sit in between the two sides, dealing on a daily basis with the referents (often beekeepers) at commune level, pest controllers and local authorities. Were they wrong? Beekeeping associations were in no doubt that spring trapping reduced predation in the apiary and this is the view of their national organisations but it's a view that could again be dismissed by scientists as 'unscientific and anecdotal'.

If the MNHN intended the release of its 'Honey bees are not the only prey' study to promote their cause over that of the FDGDON and ITSAP interim finding, its effect was quite the opposite. Here for the first time, was evidence of the damage to biodiversity caused by the Asian hornet; neither did it escape attention that the MNHN must have been sitting on the data for over ten years.

The result of the debate that followed was the first draft of a national plan for spring trapping in the vicinity of apiaries. It used far too many traps to be manageable. That number was considerably reduced in time for the 2023 National Plan but the MNHN has continued to fight its cause, intervening in the 2024 National Plan to remove spring trapping until apiaries are at medium levels of predation.

Now we should look at a quite different issue. In the rush to produce what are often expensive 'selective traps', some have made their traps too selective in respect of what goes in and unnecessarily restrictive in what is allowed out.

INRA found that whilst the Asian hornet queen will go through an 8mm dia. entrance and workers through a 7mm dia. entrance, they caught more Asian hornets by using a bigger entrance. You might have Asian hornets in your trap and no European hornets, convincing you that it's a selective trap, but if you are doing spring trapping and some Asian hornets queens choose not to go in then you have wasted your effort. Hence, a recommendation to use a larger entrance than the minimum i.e., use 9mm for the trapping of foundress queens and stop as soon as the European hornet foundresses emerge, which is later than *V. velutina* and/or by the end of the first or second week in May. You need to look at what is happening and adjust accordingly.

The other half of the problem is allowing by-catch to escape. A hornet will not pass through a hole or slot that is smaller than 5.5mm. but some manufacturers have opted out of convenience for queen excluder-sized mesh but that is far smaller than could be used. For queen trapping you could go to 6mm; some think even 6.5mm.

10.4 - Selective trapping

We can reduce the by-catch problem inherent in trapping by:

▶ Not using drowning traps. Use a sponge or wick to hold the bait. Drowning traps attract flies and over 94% of the by-catch in the MNHN photograph were Diptera.

▶ Not using yellow traps that attract more bycatch than red traps.

▶ Making sure that the bait is not attractive to bees. A mix of one third beer, one third fruit *sirop* (grenadine?) and one third white wine, works well. It's a good idea to use French *sirop* which has a very aromatic content that I have not found elsewhere.

▶ Regularly inspecting the traps and releasing any by-catch. Easier said than done. If you catch *V. velutina* do not open the trap, put the trap in a freezer for an hour, empty the trapped insects out, crush the Asian hornets and let the rest revive and fly-off. Do not rinse the trapping chamber as the hornet's pheromones will attract others.

Do not handle the traps with bare hands, wear vinyl gloves to keep human pheromones off the trap.

▶ Restricting the entrance to keep out insects larger than *V. velutina* but don't set it at the minimum. The Veto-Pharma Vespa-Catch Select offers a selectable nozzle size for the Asian hornet (7-9 mm).

▶ Allow bycatch that is smaller than the hornet to escape. Escape holes or slots need to be near the top of the trap because insects will always go up to escape; but then there is another problem because the bait's odour escapes from the trap through the escape holes. I have seen hornets crowding round the escape slots that then fly off having not found the entrance. Make sure the escape holes are close to the entrance funnel.

10.5 – Summary on Spring Trapping

No one wishes to see our biodiversity damaged by the sort of widespread indiscriminate trapping that took place in many areas of France following the arrival of the hornet but that was predominantly by the general public.

Wide-area trapping remains controversial because so much of the trapping is non-selective, has proved near impossible to control, and for that reason alone it should not be done.

However, it is widely accepted that selective spring trapping by beekeepers in the area around an apiary, for a limited period, can make a significant difference to the number of nests within predation range.

It's a wasted effort if the beekeeper uses the minimum opening on traps. The entrance needs to be 9mm and any escape holes 6mm.

Trapping selectively is not just about the trap, it is also about trapping at the right time i.e., start when the foundresses start to emerge and then stop when the foundresses are confined to the nest, the clue to which is when the first European hornets appear in the trap. Those who set a defined period for spring trapping are ignoring the vagaries of the weather and also overlook a staged emergence of foundress queens.

INSPIRATIONAL
FRENCH
BEEKEEPERS

Denis Jaffre is a Breton beekeeper who set out to devise a selective trap that would minimise by catch. His concept has been copied in many other selective traps now on the market. An active member of his department's bee health association, he has devoted many hours to educating beekeepers and promoting the use of his selective trap funnel and trap. In 2018 he was award ed the prestigious 'Grand Prix du Concours Lepine' at the 117 th Concours Lepine International inventions show held annually in Paris. He is photographe d above, with his trap and awards.

– Photograph courtesy of Denis Jaffre

11 - Defences Around the Apiary

11.0 - Introduction

The Asian hornet foundress queens establish their nests in an area and that determines where the nests are for the rest of season. Spring trapping of queens that have established nests within foraging range of an apiary can reduce predation on hives in the apiary by 50 – 80%.

11.1 - Spring Trapping

Foundress queens emerge from hibernation once the temperature is consistently above freezing and reaches circa 12 degC on 4-5 consecutive days. This can be as early as mid — late February. Initially the hornet queen will focus on restoring her hibernation losses before starting to build her first nest. That takes about a week and she will lay around a dozen eggs. The larvae take about 6 weeks to develop and emerge as the first workers.

There is therefore a window of around 8 or more weeks when the queen is foraging on her own. As foundress queens don't all emerge at once you have a window for trapping that lasts until early May. Don't trap beyond that because:

1. The foundress queens will by then be confined to the nest.
2. Too many other insects will by then be emerging, especially our own European hornet. In fact, you should stop trapping when you see the first European hornets.

The timing is fixed by the weather not the calendar so you need to observe what is happening.

Use selective traps to minimise bycatch and they must be non-drowning traps or you simply attract lots of flies (see section 10 - The Trapping Debate).

There is no point using traps that are bigger than the Veto-Pharma Vespa catch Select because you will use more bait than necessary and you need to change it every week (fortnight at most). Set the opening for 9mm.

There is no doubt that commercial baits such as Veto-Pharma's or Suterra's Trappit work well but they are expensive. The French stand by their standard cocktail of 1/3rd beer, 1/3rd white wine and 1/3rd dark fruit sirop. The beer needs to be real beer not the hoppy light stuff, the white wine is to put off your bees, and the sirop needs to be proper aromatic fruit sirop such as is made by Frucci or Teisseire and available from Sainsbury, Waitrose or online. Grenadine is a flavour recommended by the Belgians and don't use a zero-sugar type.

Graphic image of Montingac - The apiary is the yellow dot and the blue and red dots are the actual locations of nests in 2009.

The graphic shows the town of Montignac in Dordogne. A small town with a population of under 3,000, it is home to the famous Lascaux caves. Shown as red dots and precisely located are the reported Asian hornet nests as at 31 December 2009.

We will follow the protocol for trap placement as recommended in the new French 'National Fight Against the Asian Hornet'.

I have placed an apiary on the edge of town (yellow triangle) and put a 600-metre radius around it (blue dashed line) because that is a typical foraging radius from an Asian hornet nest; although the hornet may forage for well over a kilometre. In a circle with an area of roughly 1 sq. km. you can see there are 6 nests (in Brittany and Normandy 10 - 15 nests per sq. km. in a built-up area is quite normal) but this was in 2009.

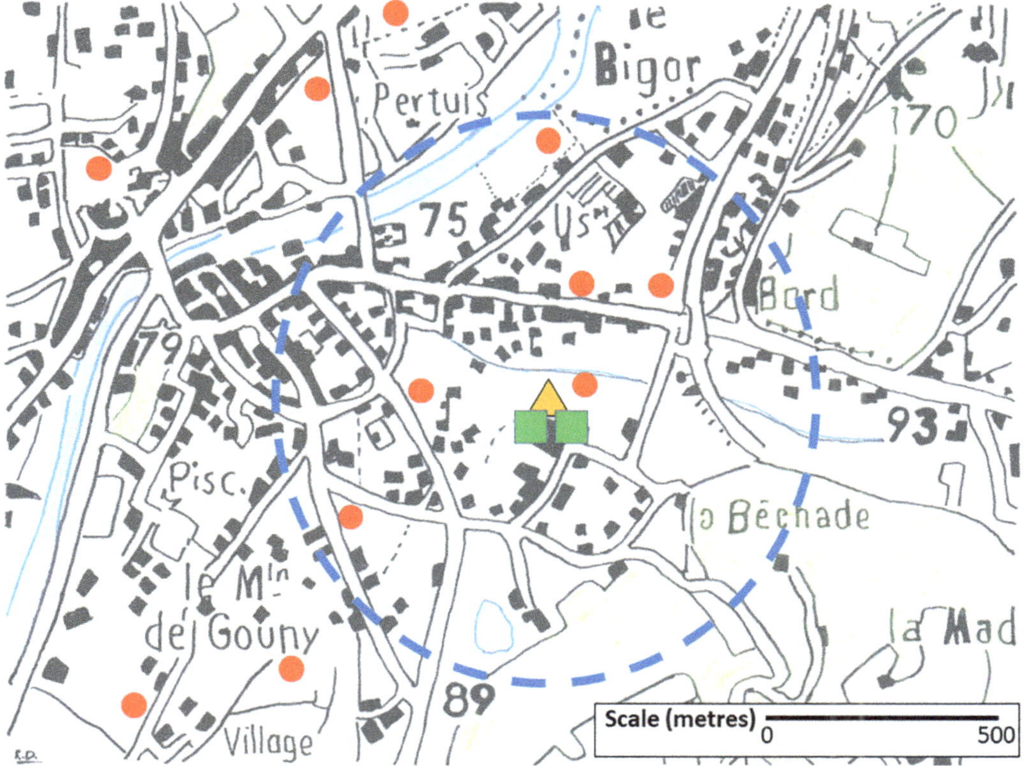

Graphic image of Montingac – Place two traps in the apiary (Green squares).

First put at least two traps in the apiary (shown as green squares). Think about where the wind is coming from because we want any foundress queens to find the traps before finding the hives.

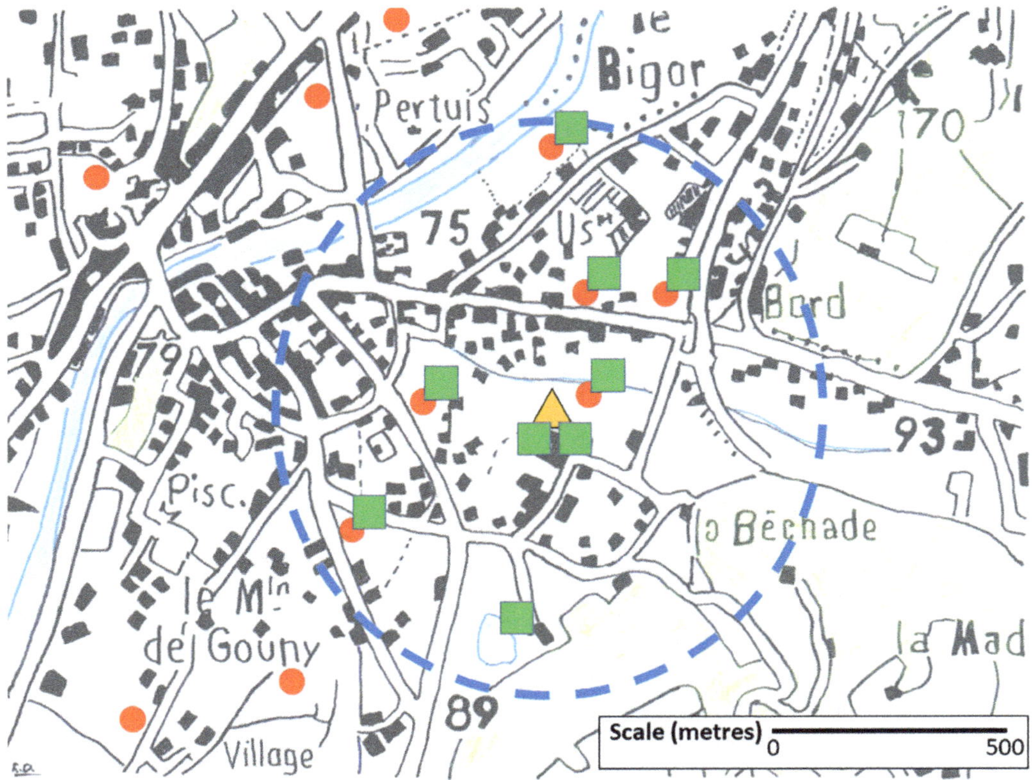

Graphic image of Montingac – More traps near last year's nests and water.

Now place traps within the radius near any known nest positions from the previous year and by any bodies of water. Hornets are like other vespids; they tend to nest where there was a nest last year and they need water, so they nest near water.

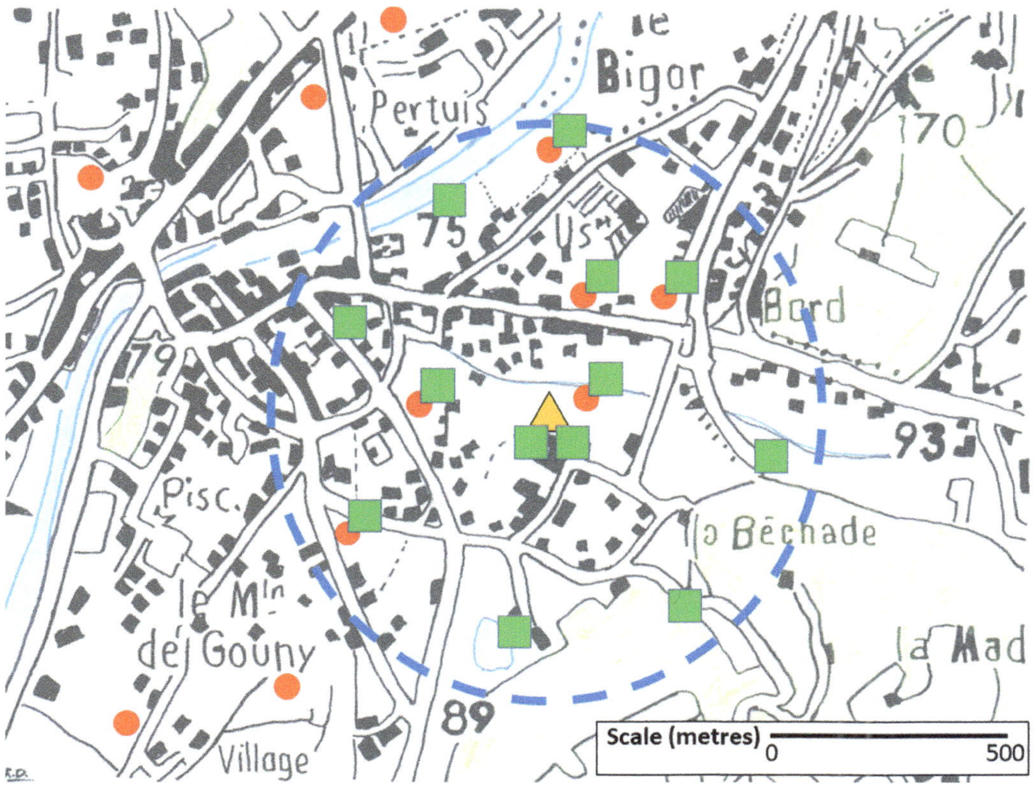

Graphic image of Montingac – Complete the pattern with traps 500m out and not more than 350m between them.

Now look at where your traps are and finish the pattern by getting a ring of traps 500m out from the apiary with no more than 350m between them. If you have to prioritise then prioritise the area downwind of the direction of the prevailing winds because those are the nests most likely to find your apiary by following olfactory clues to it, so you really want to catch those foundress queens.

You want the bait to give off its odour to attract the hornet and so the French recommend that traps be placed in the sun to warm the bait and put them at a height of around 1 ½ metres. It helps if you keep the trap small (less bait to renew) and a rain cover is essential. Renew the bait regularly, if the trap is in the sun, it will evaporate quicker than you might think. A separate bait container inside the trap warms the bait very quickly and even on cold spring days I have measured bait temperatures of 20°C or more.

I have not mentioned autumn trapping of foundress queens, which is favoured by the scientists because of the bycatch issue. It's up to the beekeeper but as 95% or more die over winter and you are really interested in the ones in your apiary's immediate area next spring, I question how sustainable it is for the beekeeper.

Do not forget to visit your traps regularly to release any bycatch. My thanks to Rachel Durham for drawing the graphics.

Photograph by the Author. Traps should be placed in the sun and kept small. As you can see, these were too big for the vapour to escape.

11.2 - Destruction of Embryo Nests

If the destruction of the hornet's secondary or main nest is a matter to be left to the professional, and many beekeepers have been injured trying to do it themselves, the location and destruction of the embryo nest (the *Nid primaire*) is a different proposition.

In French departments that have (or had) nest destruction campaigns, a lot of emphasis is placed on the public spotting and reporting embryo nests and they usually make up around 10-15% of the nests destroyed, although this can fluctuate. For example, in Morbihan (Brittany) the percentages were 2022 – 11%, 2021 – 25%, 2020 – 12%.

Always under cover and hanging down by a petiole, it is much easier to destroy an embryo nest in those first weeks when the foundress queen is on her own than to destroy an interim, secondary or main nest that will be vigorously defended by its workers.

A foundress queen and her '*nid primaire*'. She must be there when it is destroyed or she will build another. Photograph reproduced under licence – Filipe stock.adobe.com

If an embryo nest is found, it can be reported for destruction or, as long as the nest does not have workers, destroyed using one of two techniques. A foam wasp nest spray or, if the nest location is suitable and the nest is hanging from a flat surface, by placing a large jar under the nest and pushing it up to enclose the nest, then cutting the petiole by sliding a knife across the top of the jar and sealing the nest and queen in the jar. If the nest has workers do not attempt to destroy it yourself.

Whichever technique is used, it is essential to make sure that the queen is there or she will simply make another nest. This is much more likely at night and one way to check is to use a red LED backlight off a bicycle so you can see the queen but she doesn't see you (vespids are blind to red light) and tap very gently on the area near the nest. If she appears at the entrance, you know you have her.

12 - Defences in the Apiary

12.1 - Introduction

Let's be clear that there is no one solution to the Asian hornet, integrated defence involves a series of measures, each of which contributes. Furthermore, you do not need to employ all the measures outlined here until the predation level demands it.

A comprehensive array of defences in the apiary. Not all will be required until predation levels warrant it. Photograph by the author.

A question that comes up time and again is that of the cost of defences. If the beekeeper loses just one super's worth of honey, that alone would more than cover the cost of a museliere and entrance restrictor, which is all that they are likely to need initially. Lose a colony and the *harpe electrique* doesn't seem so pricey after all, especially if UK manufacturers start producing them and the price comes down.

There are four elements to defences in the apiary.

1. Apiary Environment – making the apiary work in favour of the bees.
2. Reducing the level of stress on the bees – thus delaying or avoiding foraging paralysis.
3. Reducing the level of predation in the apiary, which also helps reduce stress.
4. Defending the hive.

There are a couple of measures that I am not covering. If you have chickens and want to train them to attack hornets, that has worked in France, nor am I covering the use of squash rackets. I did hear of a beekeeper who swatted almost 2,000 hornets to add to the thousands he caught in traps. I guess it has a certain therapeutic value but was it the best use of his time?

12.2 - Apiary Environment

First, let's consider where your hives are located because the Asian hornet requires additional considerations to be taken into account on top of the normal beekeeping constraints.

Photograph by the author

Single hives or a couple of hives tucked away in the corner of the garden, like these above, are vulnerable because there's only one or two of them.

It's better to spread the pressure of predation. Furthermore, it's going to be difficult to protect those hives. They are vulnerable underneath the brood box (hornets under an OMF stresses the bees) and we can't place advanced defences such as the *harpe electrique* properly.

A much better layout is the one below. Four or five hives is thought to be a minimum grouping. Furthermore, a pair of '*harpes electrique*' can cover 4-5 hives – if the hives are spaced apart correctly.

There was a time when French beekeepers thought that hives should not be placed in a line because the hornet would hunt along the line, but researchers did not support that observation. Furthermore, we can place defences effectively and economically both in front and/or behind the hives.

Photograph by the author.

Some beekeepers think ahead and place vulnerable hives (such as nucs) in the centre of the apiary so the hornets encounter the stronger colonies first.

There are other things you can do to help your bees:

▶ The hornet is attracted to the apiary by olfactory clues. Do not draw attention to the apiary by leaving hives open, leaving out frames, extracting honey nearby — you will attract the hornets.

▸ The bees are very aware of the hornet's presence in front of the hive. They are even known to send out scouts to see if the hornets have gone. The first thing we beekeepers must do is simple – stop cutting the grass in front of the hive! The ADAAQ (Association for the Development of Apiculture in Aquitaine) recommends growing the grass long in front of the hive to give the bees a place to hide and to interfere with the hornets hawking but in the UK there are two problems that I have encountered. Firstly, the grass has often died off before autumn, which is not what you want. Secondly, if you let the grass grow right up to the hive and if the hive has CBPV, you could easily miss a pile of dead and dying bees; until you smell it! I recommend leaving a narrow strip in front of the hives.

▸ Close off the gaps between the hives and underneath to remove the hornet's hiding places from which it pops out to ambush the returning bees. Don't be too keen to cut grass under or between hive stands.

▸ As with hives that are under severe wasp attack, perhaps you can move the hives to another location where the pressure of predation is lower. Remember the hornet does not have a waggle dance and individual hornets must find the hive.

Summer Sun-shades double-up to provide screening to the rear of hives – Photograph by author

12.3 - Hive Stands & Hive Identifiers

A deep hive stand with partitioned base to keep out draughts

The hive stand in the photograph is deep and is partitioned to match the OMF floor in order to keep the winter draughts out of the hive. The partitioning of the stand makes it easy to fit a 6mm x 6mm sq. mesh underneath, thus protecting the brood box above and getting the required separation from the hornets. This solves the stress problem caused by the hornets but don't forget to seal off the back of the OMF!

Note: If your stands are not suited to this adaption there is an alternative – see section 12.4.1.

Note also that the hives have different shape and colour identifiers on them. This is a common practice across the English Channel. It enables the bees to locate their hive easier and reduces drifting (with its consequent risk of spreading of varroa and viruses). I have always used them and notice that the orientating bees use them and returning bees often go straight to the marker and drop down to the entrance.

As we tend not to mark our hives, it is useful to go over the rules for hive marking:

Colours that the bees can distinguish:

 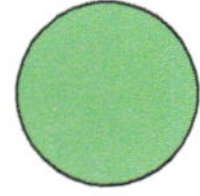  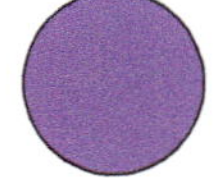

Shapes the bees can distinguish

      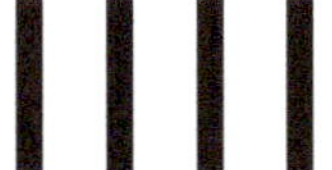

Mix the identifiers up, put them in pairs (e.g., a colour & a shape) and make sure they are different to the ones next door (bees look at adjacent hives as well as their own in order to fix their hive's location).

Lines of hives with hive markers in southern France – Photograph reproduced under licence – Mike Workman Adobe stock.com

When putting hives in pairs we tend to either orientate the hives in a different direction or put them well apart, which also means that we can get a hive tool to the boxes and keep our hands from being immediately over the bees/ frames.

In the context of the Asian hornet this creates two problems:

1. It tends to leave a wide gap that hornets can fly through to ambush bees coming in to land.
2. It makes the placement of defences such as the *harpe electrique/ arpa* very cost-inefficient in terms of the ratio of hives: *harpe/arpa*.

If you get predation in the apiary with hornets hawking the hives, close the gaps to the minimum you can manage whilst still being able to carry out hive inspections (perhaps reconfiguring over winter?).

Later versions of these hive stands were sized to allow three hives to be placed on each stand whilst still allowing enough room to get a hive tool to the boxes. A triple-hive stand makes for easier uniting and the extra space between hives makes for easier inspections, as an inverted hive roof can fit between the two hives that normally occupy the stand. It's just thinking ahead…

Triple hive stand with screened base. Photograph by the author

12.4 - Reducing the Stress of Predation

In section 8 ('The Hornet and The Honey bee') we saw that stress is the hidden killer and 'stressed' is the word used most to describe bees predated by the Asian hornet. The objective is to encourage the bees to continue foraging as long as possible. The signs that the measures are succeeding are bees foraging and an absence of a beard of bees at the entrance.

What follows are measures that reduce stress.

12.4.1 - The Open Mesh Floor

A problem raised by beekeepers is the stress caused by hornets immediately under the OMF floor. One recommendation suggested in a FNOSAD article (National Federation of Bee Health Associations) was to use solid floors but that overlooks the fact that beekeepers attribute a major cause of colony loss to varroa and the solid floor advocates may have forgotten the prime purpose of the open mesh floor.

We have already looked at hive stands that can have mesh fitted underneath to keep the hornet away from the OMF (Apiary Environment). If your hive stands are not suited to this adaption, an alternative is to use a super under the OMF and fix 6mm x 6mm mesh or a second OMF screen to the bottom of the super (E.H. Thorne – search: expanded galvanised or stainless steel mesh).

If you go down this route don't overlook the reduced area of wood that is now supporting the OMF and remember the super will not be propolised to the OMF. The danger is that they could slide apart at an awkward moment in a hive manipulation. A suggestion is to fix the super to the floor using a hive staple or fastener (see E H Thorne Hive Hardware – Moving for various fixings) or use pins/dowels as shown in the photograph below.

A low hive stand with a super and screen – note the locating dowels to stop them twisting
– Photograph by author

12.4.2 - *La Museliere*

The '*museliere*' or muzzle is an important defence against the hornet, it is a mesh screen that is proven to reduce the stress on the bees and helps maintain foraging. It was invented by Andre Lavignotte from the Pyrenees-Atlantique department.

A scientific study ('A biodiversity-friendly method to mitigate the invasive Asian hornet's impact on European honey bees' (no. 65 – see Further Reading) confirmed that the *museliere* reduces foraging paralysis by at least 40% and modelling suggested that the colony has at least a 50% better chance of survival.

Beekeepers know that it works much better than that. The problem with the scientific study was that it used 6mmx6mm gauge mesh *museliere*s that the scientists noted were restricting foraging. In addition, the estimate of increased survival was arrived at by computer modelling and that always involves a series of assumptions by the researchers. The scientists themselves concluded their study by recommending further studies using larger gauge *museliere*s!

"The following month, muzzles were crafted to keep the hornets off the board. This installation de-stressed the bees and we saw their defensive formation disappear", says Amélie Pérennou (Caen Municipal Beekeeper).

When a *museliere* is used, large clusters of bees around the entrance are greatly reduced and normal foraging can continue. It's exactly like putting a curtain wall around a castle keep but there is a bewildering array of *museliere* designs on offer. Some definitely work better than others and every beekeeper seems to have their own idea of which works best.

Let's start by understanding what it does and where the compromises lie.

Physical Separation - The close proximity of the hornet just outside the hive stresses the bees. As well as the threat of bees being captured, the hornet's pheromones stress the bees and so physical separation and distance from the hive entrance plays a role. Keep the hornet at least 200mm. away.

Ease of Passage - The hornet will attempt to capture the bee as it is coming in to land and it is known to snatch bees off the landing board. The bees are well aware of the hornet's presence and returning foragers come in at high speed, sometimes so fast that they bounce off the hive and are caught by the hornet on the rebound. If the *museliere's* screen is too small, the bee takes too long to get through and may be caught on the screen. Additionally, the original 6mm x 6mm mesh can cause pollen baskets to be knocked off, a very undesirable situation.

Defensible Zone – The bees will adopt the inside of the *museliere* and defend it by attempting to ball the hornet. Inside the *museliere* they can do this without being picked off by other hornets hawking in front of the hive.

Mesh size - A solution to the small size mesh is to use a larger mesh size that enables bees to get through the screen much quicker whilst retaining a barrier that the hornet is reluctant to cross. 13mm x 13mm mesh is now the 'standard size' but the bees find a 25mm x 25mm mesh much easier and quicker to adapt to.

Even though these mesh sizes are large enough for the hornet to get through; until hornet numbers build up in the apiary, the hornets realise that they are vulnerable inside a *museliere* and stay out of it.

But anything over 6mm x 6mm means that some hornets will go through and from September there are increasing numbers of hornets outside the hive. As the weather gets cooler the bees cluster and the deterrent of the guard bees is lost. Many a beekeeper has lost their hives in October/November by overlooking this. The answer is the entrance restrictor that we'll look at under Defending the Hive but now you know that the *museliere*, in whatever form, is a compromise.

Larger mesh *muselieres* should be put on as soon as hornets are seen hawking in front of the hive (usually July onwards) and, if the hornet is established, preferably before predation starts but if you use them, see also 'Defending the Hive – The Entrance Restrictor'.

Standard 13mm x 13mm mesh *museliere* Photograph by the author

Larger 25mmx25mm square mesh *museliere.* Photograph by the author.

E.H.Thorne *Museliere*

The E.H.Thorne *museliere* (below) has a 12mm x 23mm mesh. Note that it is made to fit a hive with frames configured cold way (take any landing board off) and in that configuration the "inset" against the front of the brood box leaves a precise 5.5mm gap so that any bees that are above the museliere can walk down to the entrance.

E.H. Thorne version of the *museliere* with a 12mm x 23mm mesh Photograph by the author. Note that this is for a hive with frames configured 'cold way'.

It can be fitted to a hive configured warm way if the hive has an E H Thorne landing board. In which case, take the screws to the landing board out and position the board and museliere carefully, setting the 5.5mm gap at the front of the hive with a drill bit. I recommend filling in the two square holes left by the "inset" fitting but it is probably not essential.

If you are making your own *museliere* consider using stainless steel mesh. It will last longer and the wire is kinder to the bees.

12.4.3 · All-mesh-*Muselieres*

You can make a *museliere* using just a sheet of 13mm x 13mm mesh (don't forget to leave lugs so that you can bend them over to join the panels together).

Use a super as a template

Cut as shown

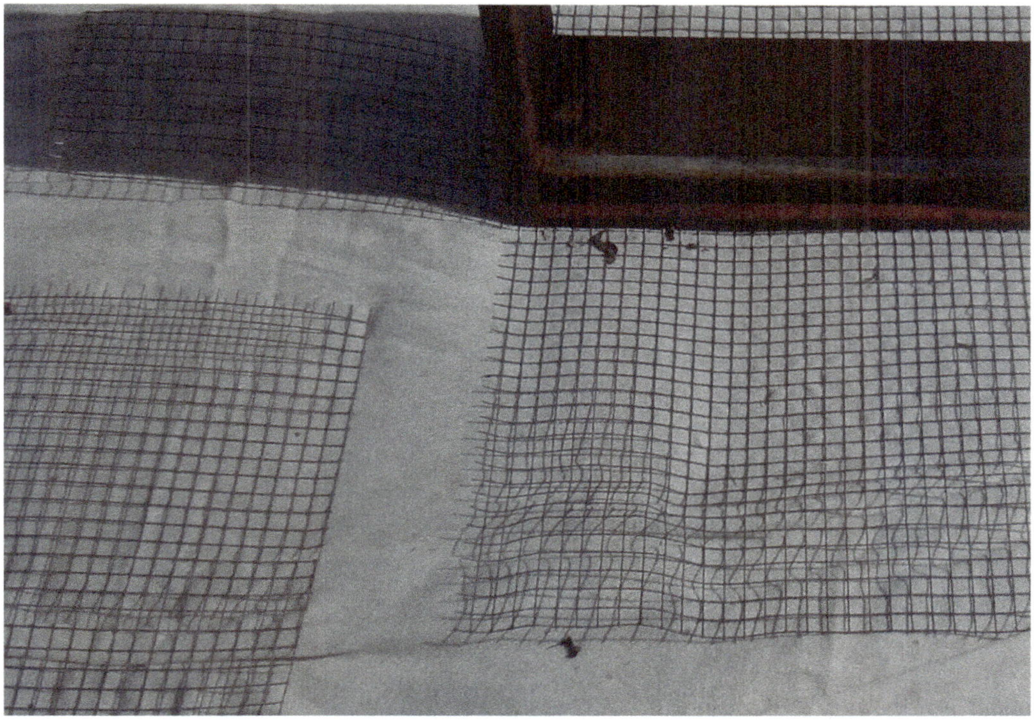

Leave lugs to join the panels.

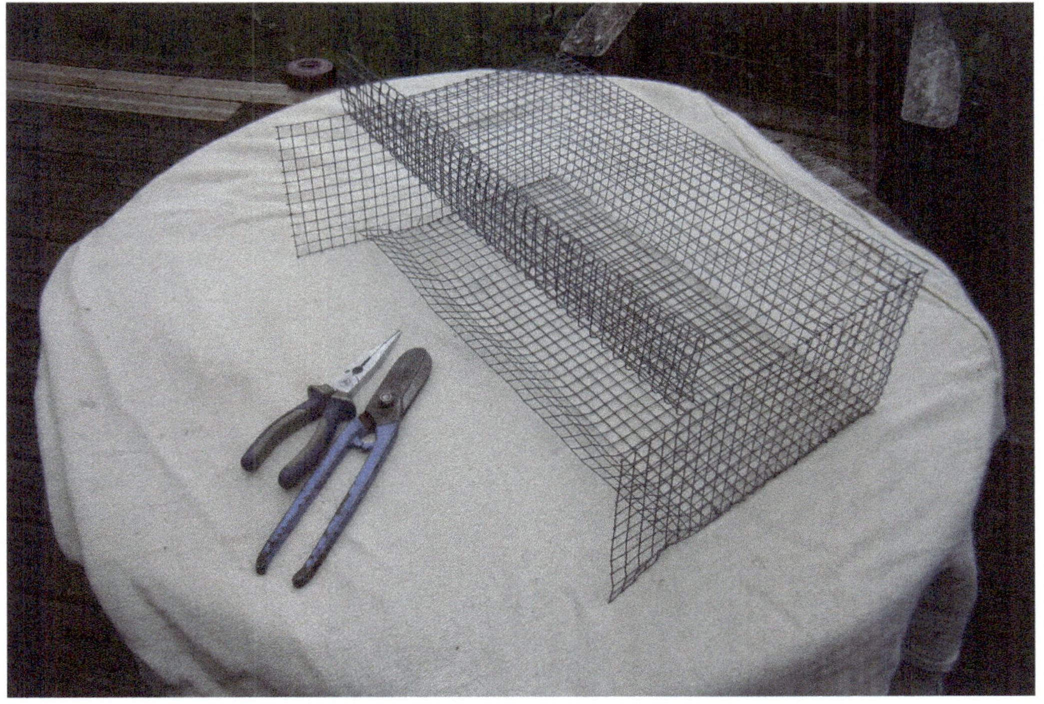

Shape to fit your hive, including any entrance board.

Staple in place.

12.4.4 - The Stop-It

The Stop-It is a variation on la museliere. It comes as a flat sheet with push-to-fit tabs and is quickly assembled before being pinned or stapled to the hive. The most notable difference to a standard museliere is that the oblique oval openings are on all five sides making it very easy for the bees to get through from a number of different directions.

Frederique Ripet, an apicultrice in the Lot et Garrone Department, thought that the wings of her bees were damaged by the mesh musliere and so she developed her *bouclier* (shield) 'Stop-IT' with larger openings that they could more easily fly through. She also set up her own company to market it. In 2024 she was award ed the prestigious Prix des entrepreneuses Helena Rubenstein et Force Femmes'. Frederique traps foundress queens in the spring but otherwise she relies solely on her bouclier 'Stop-IT'.

Opinions on it are divided. Some beekeepers say it works well but others complain that the hornet gets inside and that it confuses the bees.

> **Discussion:** I like it but I think you need to recognise its limitations. Whilst the bees fly out easily enough, those returning will initially hover in front of the Stop-It and many go behind the drop-down panel and try to get in via the landing board. That stage passes as the bees learn to fly in through the holes but it is best to put it on early to get the bees used to it. You don't want bees hovering, looking for the way-in, with hornets outside the hive.

For me, the big advantage of the Stop-It is the ease and speed of passage once the bees are used to it. The bees fly out in all directions and they can fly very fast back in to the safety of the guard bees. Furthermore, the opaque plastic that Frederique told me was not an original design intention, makes it very difficult for the hornet to track the bees inside.

Bouclier 'Stop-It' - Photograph by Frederique Ripet

I think this is an excellent choice for early/light predation, so could be a good choice early on in the hornet's establishment phase. Used with the *harpe electrique* you could have a very good combination.

Photograph by the author – 'Stop-IT' fitted to national hive with landing board.

If you use the Stop-It with a National Hive fitted with a landing board you will create a sharp angle that confuses bees still intent on using the landing board. Easily solved with a frame bar pinned underneath (see photograph below). Don't forget to use an entrance restrictor in the autumn.

Frederique has subsequently brought out the 'Stop-It Max', a standard version but with an inner screen of 25mm x 6mm slots that the hornet cannot pass through as we get into the peak predation/ clustering period. I have no experience of this product. It is more expensive than using a standard Stop-It plus an entrance restrictor.

Photograph by the author – 'Stop-IT' fitted to national hive with landing board. Frame bar prevents bees from getting trapped in the sharp corner created with a landing board.

12.4.5 - The AAVO Tube Muzzle

The Val D'Oise is a small department of only 1250 sq. km. sitting to the north of Paris but it's AAVO (Amis de Abeilles de Val D'Oise) is very proactive and as the hornet became a problem it formed a team of beekeepers to set up its own apiary as a testing ground for the whole range of apiary defences. What they had to say about predation in 2023 goes to the heart of what this guide is all about:

Les Amis des Abeilles du Val D'Oise (AAVO) – 4 May 2024

'The numbers of hornet nests was significant and the pressure on the hives was high in September-October.

Beekeepers have, for the most part, taken care to protect their apiaries and hives so that the colonies can develop and make good winter reserves (in particular with muzzles…)'

In testing different *muselieres*, they found two designs that they were prepared to recommend. One was based on the 'venetian blind' with a gap between slats of 7mm but the winner for them was the tube muzzle, which they found had the best efficiency over several seasons.

AAVO Tube Muzzle – fitted to a National hive. Photograph by the author.

This is what the AAVO have to say about the tube muzzle:

'Bees exit and enter the tubes at high speed. Hornets are very hampered in flight and do not enter the dark tubes.

The outlet will preferably be via the light-coloured tubes inclined at 45° and the return via a dark (black) tube at 90°.

It is advisable to paint the inside of the 90° elbow and also the inside of the muzzle black. The exit or white tube must be oriented towards the rising sun so that the bees memorize the direction of exit. Several models are defined for small or large colonies. The boxes are adapted to the hive models. They are dark inside. The manufacturing is simple for a beekeeper who is a bit of a handyman. The 80mm elbows can be found in DIY stores. 50mm models have also been tested with less success.

When assembling the muzzles, make sure to 'fix' the sides to the sides of the hive body so as not to leave any space that would allow the bees to escape through these small gaps.

An improvement tested in 2022 and confirmed since: put a grid (3D printed) at the base of the elbows that prevents hornets from entering the muzzle. This grid (75mm in diameter and 6.5mm passages for bees) is positioned from below from mid-September when the hornets become more intrepid...

We have noticed that the bees consider the dark muzzle to be part of the hive and clean it completely (even with the grids below). If we insert a green anti-hornet door at the door of the hive, the bees no longer clean the inside of the muzzle...'

To which I would add:

1. If you download their template, don't forget the difference in hive widths.
2. You will have to fix the tube in the top plate, work out how you are going to do that before screwing it all together!

Discussion. Once again it is a compromise. Some beekeepers are uncomfortable about not being able to see what is happening at the hive entrance for sound beekeeping reasons.

I do not like being unable to see those 3-D grids. Plus, it is more difficult to make than the classic *museliere*. I don't like the idea that my hive cannot clean itself of dead bees. I don't want to take one off to find it full of dead or near dead bees that I could have spotted much earlier. You'll have to make up your own mind on this one but don't overlook the risk that in solving one problem, we can create another that we may not foresee.

12.4.6 - The Cabane Grillagee or Netted Enclosure

In my research I was unable to get a consensus on netted hives although UNAF stated that they can work satisfactorily to reduce predation (UNAF Special Edition of 'Frelon Asiatique - 12 ans après son installation, état des lieux et méthodes de lutte') (no.5 see Further Reading).

The problem is that there are so many versions and no clear agreement on the size of net to use. I have seen an apiary placed under the equivalent of a tennis-court-sized fruit cage with fine (6mm x 6mm) square mesh. The idea being that the area of net is so large that the restricted mesh-size is not a problem and I have also seen 6 hives housed inside a shed-sized enclosure with similar-sized nets dropped down in front of the hives when predation starts.

An article in the November 2024 SNA magazine '*L'Abeillle*' reported that the equivalent of a large cloche with a plastic net that had hexagon shaped holes (each side of the hexagon 12-13mm, giving a height of 15mm and width at the widest point of 29mm) worked 'really well'.

The point to watch here is that if the hornet can get through the net and the net is not close enough to the hive and its guard bees, then the hornet will just hawk inside the net. Whatever you do, you must put entrance restrictors on the hives once predation reaches peak levels and the bees have started to cluster.

12.4.7 - Summary on Muselieres and Recommendations

There is a bewildering choice when it comes to muselieres, so let's sum up on the choices:

▶ The 'classic' museliere (section 12.4.3) as invented by Andre Lavignotte is the most widely used. Don't bother with the 6mm x 6mm mesh versions, you can achieve a better result with 13mm x 13mm and I particularly like E H Thorne's 13mm x 23mm mesh version. But don't forget the entrance restrictor.

▶ The all-mesh museliere (section 12.4.3) is a real economy job, suggested by a French beekeeping association. I have doubts about that one but better than not using a museliere at all.

▶ The Bouclier 'Stop-IT' (12.4.4). As I said, a mixed response from users and it is important to get the bees used to it. Having said that, my bees rushed out in all directions and soon learnt to come in the front. It doesn't hinder them at all, so I am going to use mine in a low-medium predation scenario and, because I have harps, it may not be necessary to move to the classic at all. If the hornets get in and I have to use a classic then it will be the E H Thorne one.

▶ The AAVO Tube Muzzle (12.4.5). Came out a clear winner in the AAVO's trials. It's more expensive and complicated to build than a classic (at £4 each per pipe bend, plus pipe, plus wood, hole drill etc.) and I really don't like not being able to see the landing board or the idea of hidden grids. I am going to pass on that. The situation will have to get pretty bad before I'd want to use it.

▶ The 'cabane grillagee' or netted enclosure. UNAF found it satisfactory and some beekeepers like them. Do I want to have to move one to get in the hive? I remember a couple of beekeepers who tried a plastic net enclosure but found the hornets climbed in and hawked inside, then dropped their booty on the way out…

Remember, the *museliere* reduces stress but not the level of predation, trapping does not reduce stress.

12.5 - Reducing the Level of Predation in the Apiary

12.5.1 - Introduction

When predation in the apiary becomes a problem and the *museliere* is not enough, the beekeeper must act to reduce the level of predation, which at the same time will also help to reduce stress. They can do this in one of two ways, by using baited traps or using the *'harpe electrique'*.

A school party behind the hives, harpes electrique in front! Photograph reproduced under licence – Buddha Adobe stock.com

Baited traps are considered by some to be counterproductive because the traps themselves may attract more hornets to the apiary than would otherwise find it. INRAE removed baited traps from their apiaries because they felt that this was the case. SANVE in Spain argue strongly not to use baited traps in an apiary.

Given the cost of the alternative i.e., the *harpe electrique*, for many there is no choice but to use baited traps. When using traps in the apiary there are two points to bear in mind:

1. Hornets attract hornets, so if you empty a trap do not rinse it out and do not rush to empty a trap of live *V.velutina*, they are the best bait.
2. Handle traps with rubber gloves, don't leave your pheromones on the trap.

Although beekeepers should always use selective traps outside the apiary and for spring trapping, the argument for using selective traps (see Spring

Trapping) near hives in the main predation period (i.e., September – November) is not the same because the area around the hives becomes a killing ground for the Asian hornet.

Veto-Pharma Vespa Catch trap (non-selective) – not the best to use in the predation period, it doesn't have the capacity.

12.5.2 – The Tap-Trap and similar

The sight of a line of Veto-Pharma Vespa Catch traps placed one by each hive merely indicates that the beekeeper hasn't grasped the scale of the response required because however well they work for spring trapping, such small capacity traps fill up too quickly and they take too much management.

Too many traps requiring too much maintenance and expensive bait is not sustainable for the beekeeper. We must therefore use a trap that has the capacity to deal with large numbers of hornets, doesn't require the same degree of maintenance and one that can use a very effective bait that every beekeeper has in abundance and doesn't cost them a penny. If it also happens to be very selective, then it is an obvious choice. Such a trap exists, it is called the system BCPA or Jaberprode.

12.5.3 The Jaberprode (www.jaberprode.fr)

This is 'the' selective trap to use in the apiary during the predation window. A 2023 FNOSAD survey of GDSA's found that 30% of the traps in use by beekeepers were using Jaberprode trap modules.

With its selective funnels that incorporate escape holes, it has the capacity to hold the potentially very large numbers of hornets predating in the apiary and it uses bait that every beekeeper has available.

There have been many inferior imitations of this trap but this is the original and it was invented by Denis Jaffre, a beekeeper from Brittany, who one year lost 35 of his hives to the hornet and set out to devise a trap that would not only be effective but also selective.

His plan was to devise a cheap and mass-produced trap funnel that had a restricted entrance to avoid trapping prey larger than the Asian hornet and escape holes that would allow easy escape for bycatch smaller than the hornet. After a lot of work and some initial teething problems with manufacturing tolerances, he succeeded in getting the specification he wanted. The result was the Jaberprode funnel.

His funnels are sold separately in pairs (you may find them available in the UK, search Jaberprode) and they can be mounted by the beekeeper in a spare super fitted with an OMF screen, placed above a defunct hive that has old brood frames and frames of granulated honey.

Photographs courtesy of Denis Jaffre: Prototype Jaberprode funnels fitted into a spare super fitted with an OMF, mounted on top of a defunct hive and (right) an early prototype.

The super will need a lid and an OMF floor screen to contain the hornets.

Alternatively, it can be placed above a super/eke containing wet wax cappings in a tray but make sure the honey has fermented by placing it in a warm place (such as an airing cupboard) to make the bait unattractive to bees.

Jaberprode sell a complete BCPA trap with funnels mounted and beekeepers report that it is very effective, catching large numbers of hornets. In its complete form it is not cheap (c. 80 euros). In the UK you may get them for c.£110. Make sure that you buy the latest version, which has a plastic screen on the inside of each funnel entrance.

Denis says two traps per apiary should be sufficient, placed with prevailing winds in mind i.e., put the traps on the downwind side so that hornets travelling upwind following olfactory clues find them first and don't put them close to hives or you will lead the hornets straight to the bees.

Photograph courtesy of Denis Jaffre: Jaberprode funnels fitted into a bespoke container with an OMF, note the latest version has an inner screen fitted to the end of the funnel on the inside.

12.5.4 - A Homemade Equivalent?

It is possible to make a homemade equivalent of Denis Jaffre's BCPA trap. It is cheaper than importing one from France with the shipping and extra tax. It is not the same as the original, so if you were going over to France you should buy the original. If you can get the Jaberprode funnels in the UK it will be much cheaper than the sieve version described below, which is really for those determined to make their own trap and who cannot buy the original Jaberprode funnels.

Photograph by the author

Assemble a National brood box (c. £40) but instead of inserting the side panels (assuming 'warm way' orientation) cut them down to allow a slot at each end for the queen excluder (E.H.Thorne £6) and fix the cut-down side panels to the front/back panels.

Photograph by the author

Turn the brood box over and fit an expanded metal galvanised sheet (E.H.Thorne £10).

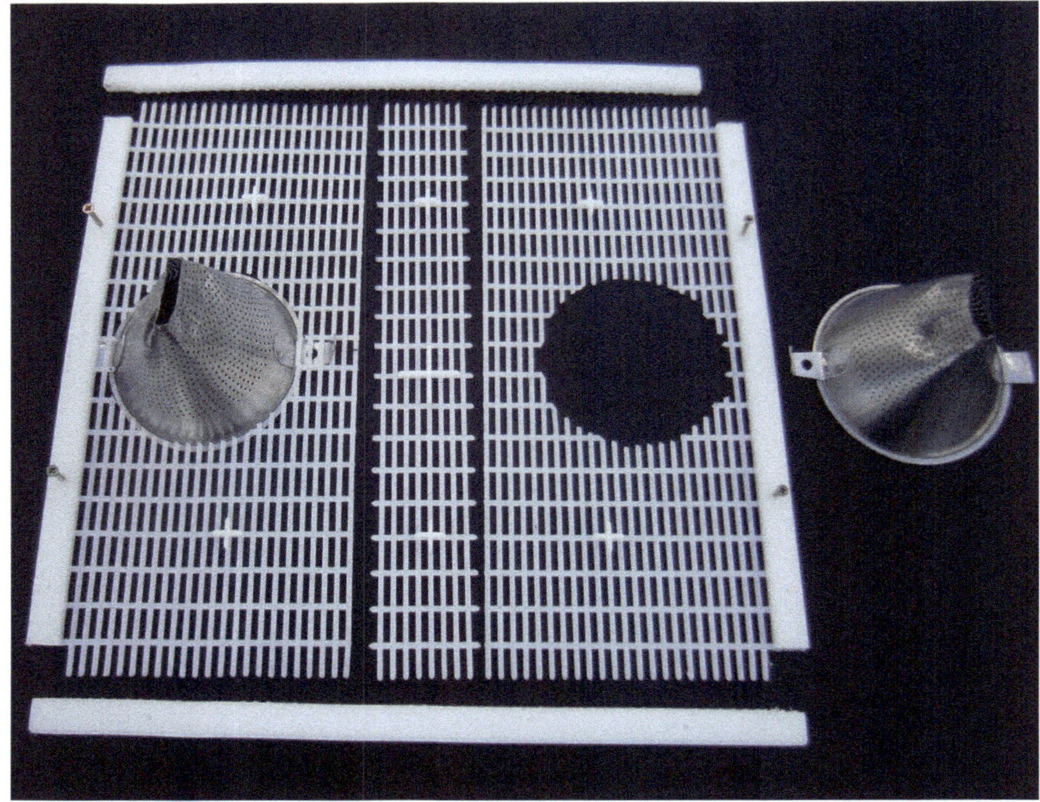

Cut the excluder exactly as shown for a National brood box. Photograph by the author

Take an E. H. Thorne plastic queen excluder and cut it as shown, with two circular openings, each cut to accept a stainless steel catering sieve.

Take a stainless steel conical sieve, flatten the end and cut off the point to make an 8mm slot. Be very careful as the edges will be razor-sharp. If you can get them, the Jaberprode funnels will obviously make the whole trap much cheaper.

Slide the excluder as shown and fix to the side rails as shown below.

If you have bought the Jaberprode funnels you won't need the excluder at all and some think that works better because too big an area of excluder distracts hornets away from the funnel entrance. Just build the box as normal and cut openings for the funnels.

Photograph by the author

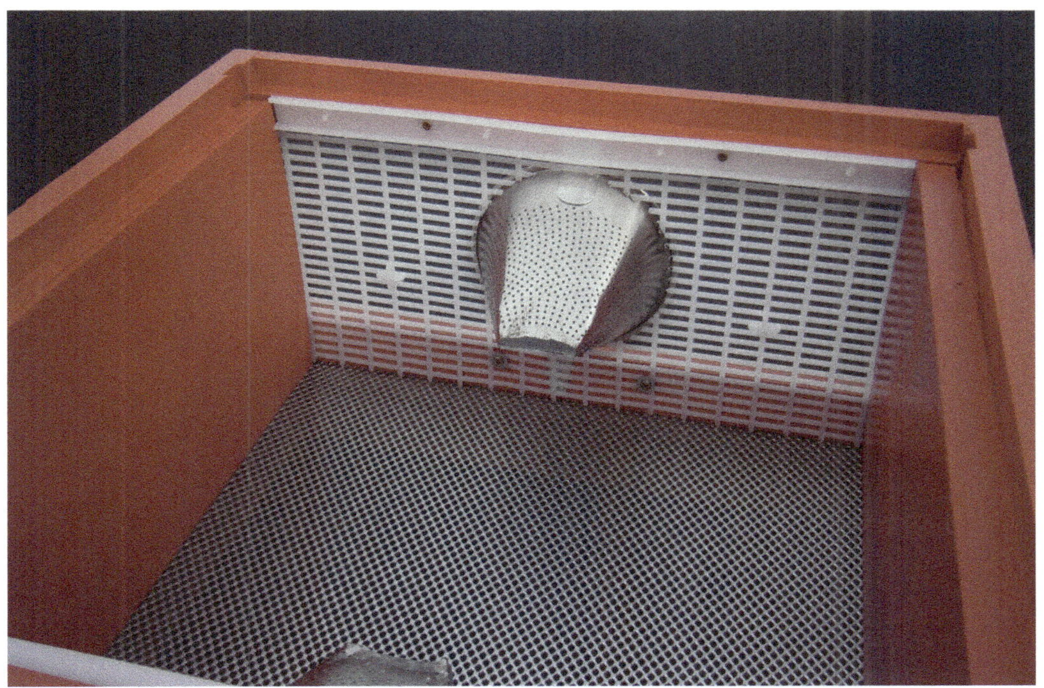

Photograph by the author

Photograph by the author

You'll need a Perspex sheet top and an eke under the OMF floor to hold a bait tray. Incidentally, the colour red attracts less by-catch.

12.6 The *Harpe electrique / Arpa Electrica*

We now look in detail at what is proven to be one of the most effective measures for reducing the level of the Asian hornet's predation on the hive; the *Harpe electrique* as it is known in France or the *Arpa Electrica* as it is called in Spain.

It is an expensive item that can cost as much as much as £300 (fully assembled with its own solar panel) or it can be homemade for around £120, and there are a bewildering number of choices/options, so let's go through the decision process step-by-step (the Annex has all the detail on making your own).

12.6.1 - What is it, Why Use it and Does it Work?

The harp is a grid of fine wires strung in a rectangular frame; the wires are electrified by a high voltage (HV) charge that stuns the hornet causing it to fall into a water bath or trap below. The hornet does not see the wires, which are sufficiently distanced from each other (typically a 20mm gap) that a honey bee passes through unharmed. The grid is wired with alternate polarity across the grid (positive-negative-positive-negative etc.).

Image of author's Harpes Electrique and Arpas Electrica with three covering a 5-metre run, a ratio more commonly found in scientific studies. All have an individual HV power unit for the grid - Photograph by the author.

You may catch some bees, especially with the wet *harpe / arpa* (see – 'Trapping chamber or Water bath or just the grid?'). Although you can reduce this by-catch problem, you cannot eliminate it entirely but the bycatch number is small and is outweighed by the number of bees that you would otherwise lose through predation.

The selectivity, efficiency and effectiveness of the electric harp is now beyond doubt but it is important to note that some of the studies used a higher ratio of harps to hives than might be the case in most apiaries. Furthermore, the researchers may not have taken in account all the factors that would be obvious to the beekeeper; e.g., the increased risk to the bees during swarming or making sure the bees have access to water and have found that to be a reliable source, so that they don't need to go to the harp's water bath, which should have had lemon washing up liquid put in it anyway.

INRAE has been conducting trials in their apiary on the Bordeaux campus, the interim results of which were reported in the SNA's magazine L'Abeille (see inset box and no.68 in Further Reading).

Article in L'Abeille January 2024 on INRAE's test of *harpe electrique* in their apiary at Bordeaux:

'Quite spectacular results'

'Besides the impressive results… this trapping reduced predation in front of the hives and their weight increased significantly compared to those not protected'

Denis Thiery, Zoe Tourrain, Monica Doblas-Bajo, Michel Costa, Gaetane Le Provost

Most beekeepers will initially purchase the *museliere* and entrance restrictor, only buying the harp when or if the hornet gets fully established and predation levels warrant it. If the beekeeper already has these then the harp might only be deployed when predation levels merit their use in late August or autumn. But that misses an opportunity to catch early hornet workers and as SANVE point out, a worker caught in July is worth 100 in August because it is only early on that catching hornet workers can seriously affect the development of the hornet nest and a hornet nest that is not fully developed cannot produce as many foundress queens in the autumn as one that is.

12.6.2 - Placement and Ratio of Electric Harps to Hives

Manufacturers of harps advise a ratio of 1 harp to 4-5 hives in a 5-metre line plus a second harp at the end, with the harps placed in front and perpendicular to the hive line. Such a layout is shown in the graphic below.

I have called this 'in front of the hive line' arrangement 'Spanish style' because the layout of the harps in this way is more commonly found in Spain. The manufacturers' recommendation is a generalisation and the beekeeper needs to take into account hive spacing and also when the harp is deployed because 'Spanish style' is not recommended if there is a risk of swarming. It also depends on whether the beekeeper is going to use the *museliere* as well as the harp. If you are not using the museliere, the harps have to go out as soon as predation starts.

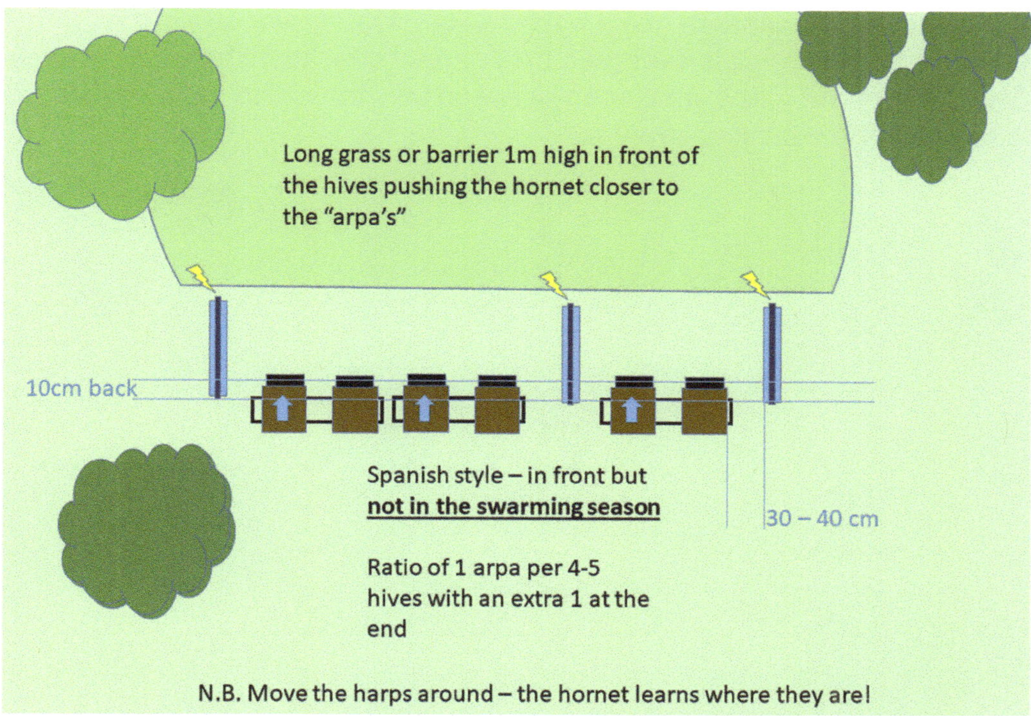

Graphic showing 'Spanish style' layout of harps placed perpendicular to hive and in front of hive line. The solid black boxes are *muselieres*.

Photograph by the Author showing two harps placed 'Spanish style' in a 5m run of hives. Two SANVE CEBO arpas in foreground. Note the central power supply with solar panels.

If the beekeeper wants to use the harp to catch early workers but is unsure that swarming is over then they should consider what I call 'French Style', where the harps are placed between and/or behind the hives, as it is safer for the bees (see graphic).

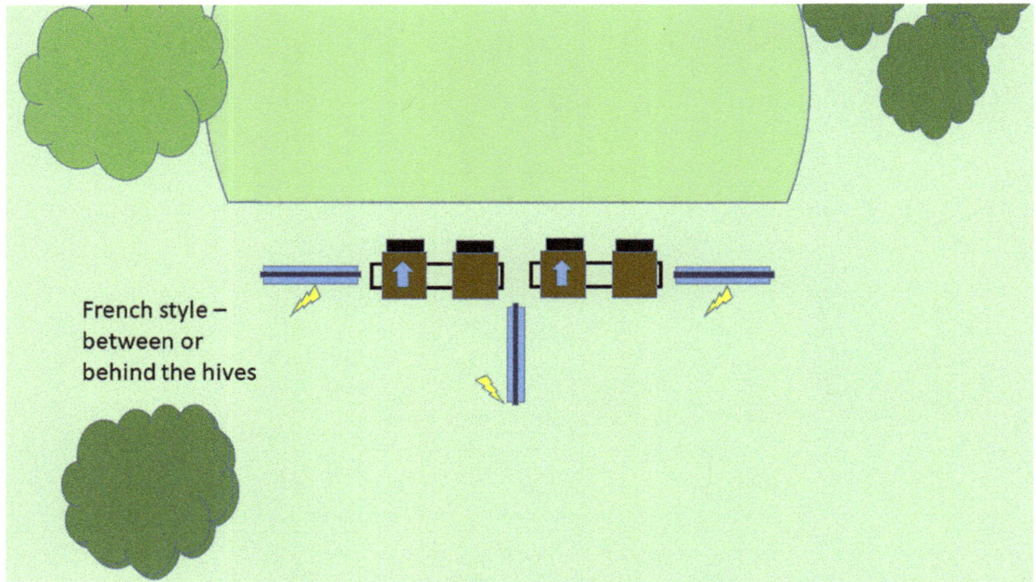

Graphic showing 'French style' layout of harps placed beside and between hives.

The Spanish style arrangement is even more effective if a barrier is placed in front of the hives just beyond the harps so that the hornet is channelled towards them. Furthermore, the hive spacing can be closed up (enabling a greater harp to hive ratio) and the gaps at the rear of the hives closed off to prevent the hornet taking shortcuts between them.

Be guided by the designer's rule of thumb; if it doesn't look right, it's probably not going to work. Look at your hives, look where you are planning to put the harps. Picture hornets flying around in front of the hives. What are the chances of them flying into the grids? If the spacing looks too much, it probably is. But don't make the opposite mistake and put a harp closer than 30cm to a hive entrance; that's too close. In fact, 40-50cm is better if you can do it.

One other point to note. I hung an HV unit on the back of a hive for convenience and within 40 minutes the hive had swarmed. That was in late September and when I put my ear to the unit even I could hear the noise coming out of the electronics. A lesson learnt the hard way and yes, that is purely anecdotal, no science whatsoever!

Hives with harps placed French style. Photograph by author

Remember that the harp is not a stand-alone solution, it needs to be deployed alongside other devices such as the '*museliere*' and the 5.5mm hornet entrance restrictor, but it may replace bait traps in the apiary unless predation levels are very severe. Spanish experts would argue that replacing bait trapping entirely, reduces the apiary's olfactory signature, so not attracting hornets that might otherwise have not found the apiary at all. Furthermore, it should part of an integrated defence against the hornet that includes good beekeeping, the spring trapping of foundress queens, the identification and destruction of the hornet's first nest ('*nid primaire*') and the reporting and destruction (by experts) of the hornet's main or secondary nests.

12.6.3 - Choices and Options for the Electric Harp

Do not be tempted to use an electric stock fence unit. It is far too powerful (i.e., dangerous) and it has a pulsed output; you want a constant output.

Assuming that the beekeeper has decided to go ahead with the electric harp, let's now look at what can be a bewildering range of choices and options. We'll look at what is available as a manufactured item.

12.6.4 - Choosing the Highest Voltage is a Mistake.

The voltage used on the harp's grid varies between manufacturers and this is the first element of the specification that needs careful consideration. The first harps were French and voltages used were in the many thousands of volts. One adjustable-voltage HV (high voltage) power unit that I own delivers between 10,000 – 15,000 volts albeit at a low current. Others generated half that voltage whilst Spanish units tended to use an even lower voltage, circa 1,500 volts.

A higher voltage unit may appear attractive to the user but the higher the voltage, the larger and more expensive the HV unit becomes, taking more power and introducing the likelihood of leaking current through the insulation of the HV feed wires or with the grid itself arcing out, thus rendering the harp useless.

The Spanish argue that a high voltage is not necessary to stun the hornet (paralyzing its flight muscles and causing it to fall into a trap or water bath). Furthermore, a lower voltage opens the door to smaller capacity batteries and to solar power.

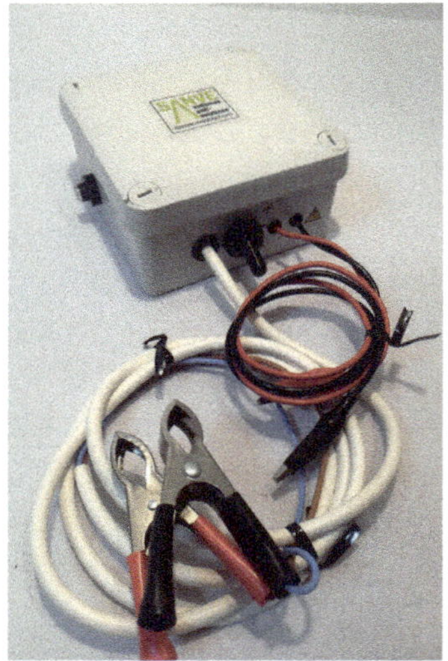

A 12V/230V French HV unit, variable HV output (left) A SANVE HV unit with 12V supply (right)   Photograph by the author and SANVE.

12.6.5 - Trapping Chamber or Water Bath or just the Grid?

Spanish *Arpas Electrica* by SANVE (SANVE.weebly.com) Dry *arpa* left and wet *arpa CEBO* right). The CEBO feature is a protected design. Photograph courtesy or SANVE.

When the high voltage grid stuns the hornet, it either falls to the ground or clings to the wire. It may recover to fly off and there is a school of thought that even this is effective because the hornet learns to avoid the area, which is why some recommend that the position of the *harpe electrique / arpa electrica* is changed from time to time. Of course, such hornets remain free to return or to predate outside the apiary and so most users incorporate either a water bath or a trapping chamber. The idea is to kill the hornet not scare it.

The water bath is effective at killing the hornet, especially if a shot of washing up liquid is added (it reduces the surface tension that might otherwise allow the hornet to float on top). However, there may be an issue if the unit is placed in front of the hive line and the bees decide to use it as a water supply. Some beekeepers have solved this problem by scenting the water with lemon juice.

Instead of a water bath some units incorporate a dry trapping chamber from which any by-catch insects inadvertently caught by the grid may escape. The dry *'arpa'* however, leaves the beekeeper with the problem of disposing of the live hornets caught in the trap. The answer is to put the trap's capture bottle into a freezer for 60 minutes, then sort out any bycatch and crush the Asian hornets before they recover.

The dry *'arpa'* will really come into its own once we have a poison approved for use as a 'Trojan bait'.

12.6.6 - Derivatives of the *Harpe electrique / Arpa Electrica*

Some harps enclose a bait tray and are used for spring trapping.

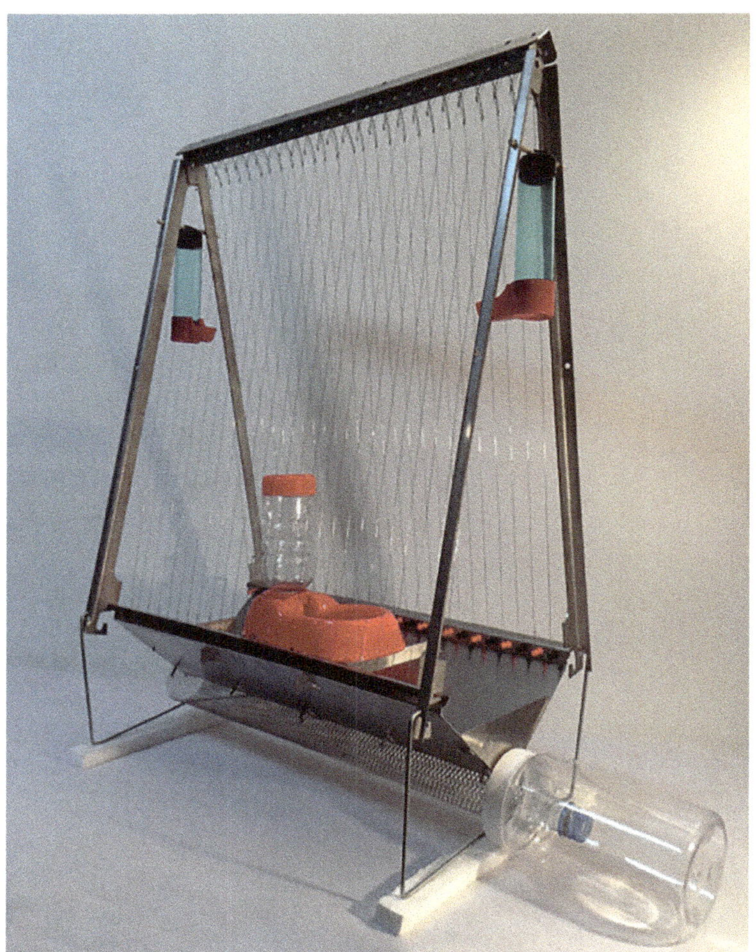

An enclosed bait trap for spring trapping – Photograph courtesy of SANVE

SANVE spotted the fact that the hornet's vision is so poor on definition that it will mistake a stationary dot of a certain size on a plain background for an insect and will attack the dot. Place a pair of electrodes either side of the dot and the hornet will continue attacking the dot until it contacts both electrodes. They call it a 'Diana Trap'.

SANVE Diana traps – Photographs courtesy of SANVE

12.6.7 - The SANVE *Arpa* CEBO

SANVE took the Diana trap idea one step further, the *arpa* 'CEBO'.

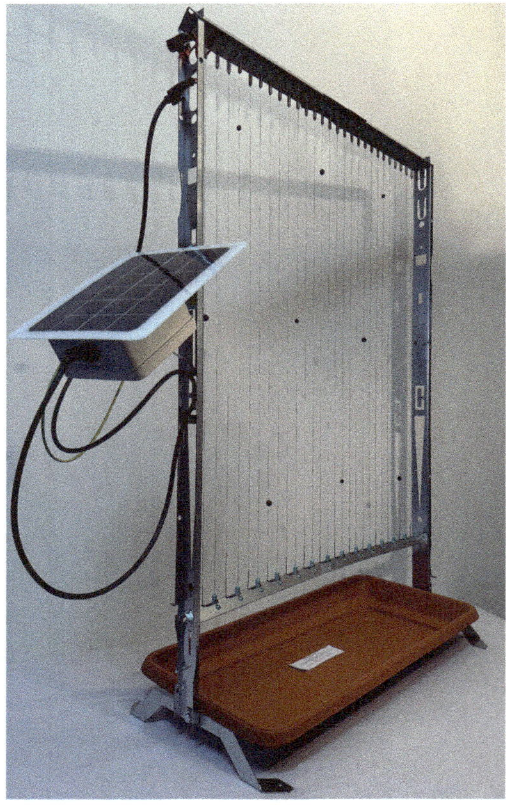

SANVE Arpa CEBO (a Protected Design) Photograph courtesy of SANVE

The arpa CEBO uses precisely sized balls that the hornets attack. It is claimed that it increases the already impressive catch by as much as 80%.

Hornets in water bath under a SANVE arpa – Photograph SANVE

12.6.8 – *Museliere* with Integrated *Harpe*

There are suggestions for a *museliere* with an integrated *harpe electrique.* Such a proposal is being investigated by the FNGTA, a French national apicultural technique advisory group. It could take some time before the outcome of their investigation is known.

I am struggling to see the benefit of such a device. The *museliere* provides a defensible barrier that keeps the hornet away from the hive entrance and reduces the stress on the bees. The *harpe* reduces the level of predation in front of the hive and gives the bees a much better chance of survival when leaving and returning to the hive. Thus, reducing the number of stop signals that ultimately results in foraging paralysis.

The hornet hawks outside the hive and has no reason to enter the confined space of a *museliere*, in fact the whole rationale of the larger mesh *muselieres* depends on it. Therefore, what does an electrified *museliere* achieve?

12.6.9 - Powering the *Harpe/Arpa* – More Choices

The *harpe* or *arpa* requires a high voltage (HV) power unit to supply high voltage for the grid and unless it is 230V version, it also needs a low voltage feed to that power unit. There a number of options:

1. Harp with a 12-volt feed to the HV power unit.

2. Harp with a USB feed to the HV power unit.

3. Harp with its own power unit, internal battery and solar panel.

4. Harp with no power unit piggy-backed off the HV power unit of another.

It comes down to the number of harps and cost. The power unit can cost as much as the harp and a power unit can supply the high voltage to a number of harps. In theory, as the harp draws no current until an insect shorts out the grid, you could have as many harps as you like fed off one power unit but now the laws of electricity intervene.

Long runs of HV cables between harps have a resistance that reduces voltage and current. The insulation on most electrical wire is not designed for high voltage, which can leak away. I prefer single-core wire for HV cables and keep the wires away from each other and keep the positive feed wire off the ground.

The more connections the more likely there is to be a 'high resistance' joint that does not allow any current to pass. In addition, if an insect gets stuck on the grid on one harp, shorting it out, the other harps are useless. You could be forever having to check the grids and clear the obstructions. Perhaps you might link two harps in tandem but think carefully which ones or you might leave a hive with no protection at all.

In short, I think individual HV power units are a much better option and I know one manufacturer is moving to that basis with their harps. Harps with an integrated HV unit are the way forward as the feed wires can be kept very short (car manufacturers went down this route many years ago with spark plug caps that incorporate the HV coil).

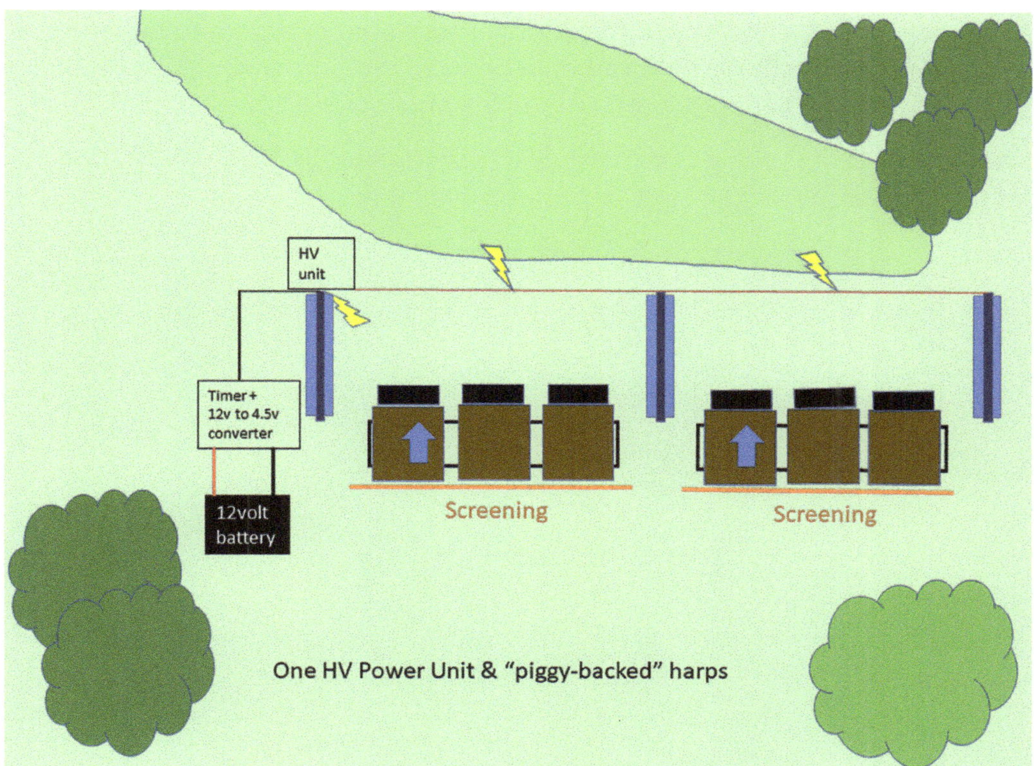

12volt battery, timer or photocell, 12V to 4.5V converter (optional), single HV unit serving 3 harps

If you are only going to deploy 2 or 3 power units you may only need a 12-volt battery that you recharge every week, assuming you have a timer or photocell; and you should have one because you don't want to have the harps on at night catching moths and bats. If you buy a power unit with a couple of crocodile clips to attach to battery terminals, make sure you get the polarity right or you'll destroy the power unit. Make sure any photocell switches off at night, not on!

A better option is to have a power unit with a solar panel but in the UK it has to have a battery because they don't work if the sun is obscured by cloud. Don't buy one that has a panel less than 20 watts because some units designed for sunny Spain won't be powerful enough to power the harps and recharge the battery in our less sunny climate (see also Section 12.6.10).

If you have a large apiary you may find it more cost effective to buy harps with HV power units that are fed off a central, more powerful, solar panel unit with a larger capacity battery (see graphic below).

I recommend that you go onto the SANVE website (SANVE.weebly.com) if only to view the large range of options available.

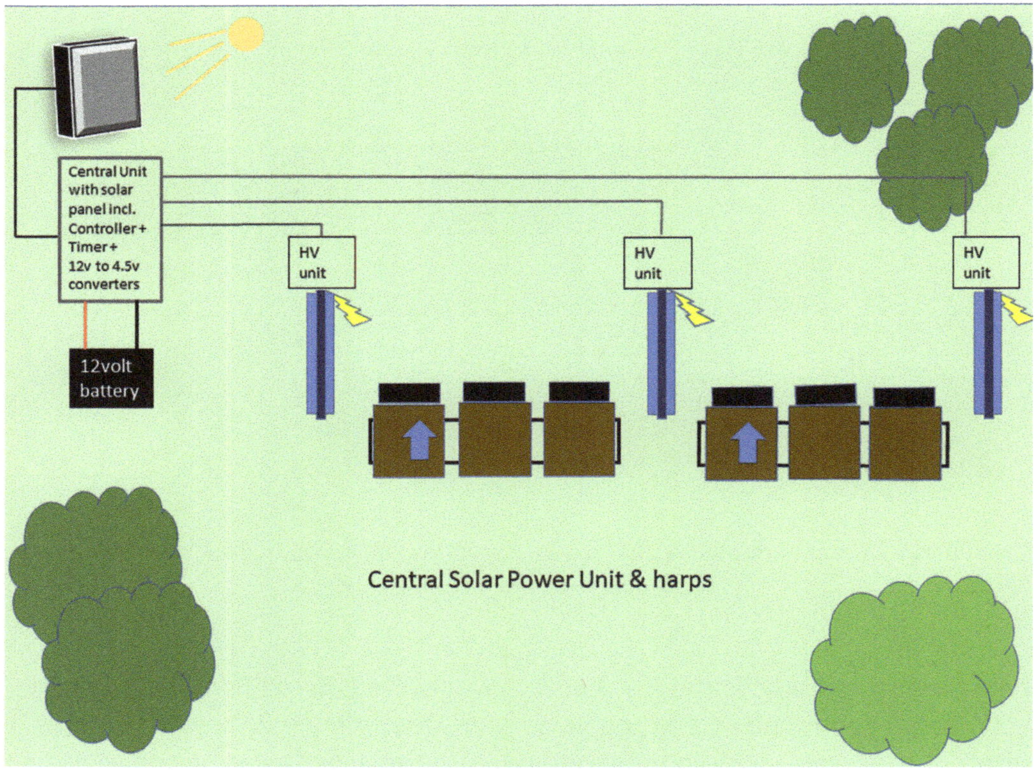

Layout of central power supply with solar panel feeding HV Units

WARNING

The electric harp presents a risk of electric shock. It is up to the reader to carry out their own risk assessment, taking into account where the harp(s) will be placed and the potential risk to both themselves and others. Appropriate barriers and warning signs should be used to prevent injury to those who might come into contact with the harp or its power unit.

12.6.10 - Assembled or Flat Pack or Building Your own *Harpe electrique / Arpa Electrica* – System – The Pros and Cons.

Let's consider the pros and cons of building your own *harpe electrique / arpa electrica*. Some will have little choice but to build their own and in the Annex I have given all the necessary detail to enable the beekeeper to design their own *harpe* and power supply system. But you will need a workshop/garage, a comprehensive toolkit, and some electrical skills to do the wiring. Whilst not difficult, it is time-consuming and you may wish to consider if it is cost effective in terms of your time and effort.

Photograph courtesy of E H Thorne. This harp is supplied with 20-watt solar panel with battery pack. This photograph is in a number of books but who made it and what size is the battery?

The advantage of buying a manufactured harp is that the manufacturer should have sorted any design issues and can use bespoke parts for which the amateur has to find a near-equivalent that may not do the job as well. Furthermore, the manufacturer has the advantage of scale. If you are buying from an established supplier, such as E H Thorne, then there is an added assurance but I must sound a note of caution.

Even though the *harpe electrique* has been around for many years, in France it is only recently that it has become the must-have item in the apiary and prior to that many harps in France were (still are) homemade. So, there has been a rush to fill a gap in the market. I recommend you check that the manufacturer (not the seller) has a history in the field and if you are going to invest in something as expensive as a harp, there are some points to look out for before parting with your money.

First of all, it is likely in this country that the manufactured harp will have been bought in from France or Spain. If it comes with a solar panel it has to have a battery (I look at this in more detail later). Whilst a 20-watt solar panel may be adequate, it all depends on the capacity of battery to which it is married because, especially in autumn, you may not get useable solar energy in the UK for days on end, much longer than would be the case in France or Spain and the panel has not only got to power the harp's HV unit, it has to recharge the battery (see – Annex - Photovoltaic panels and batteries – doing the maths.). The safest solution is one where the battery is separate, so you can see the battery's capacity and increase it if necessary.

Secondly, sound electrical connections are essential for a harp to work properly. Bi-metal reaction can create high-resistance joints. If you see different metals (e.g., aluminium to steel) in a connection exposed to the air and moisture then corrosion is going to be a problem. In addition, in our damp and often wet climate the cable connections need to be waterproof so water cannot short them out. I'd look for the type of connectors you see in cars i.e., designed to deal with wet conditions.

Watch out for 2-core HV feed wires. The insulation on most wiring is for up to 400V, so put two HV wires together in a 2-core cable and you are likely to get leakage.

12.6.11 - Fully Assembled or Flat-pack?

Fully-assembled is ready to go but obviously costs more so the real question is what is involved in flat-pack self-assembly? A SANVE *arpa* comes as shown in the photograph below. Before you put it together you need to consider whether you want to paint it. The frame is galvanised metal, which will inevitably deteriorate in our climate and do you really want shiny metal in your apiary? A can of primer spray and a large can of camouflage green from Halfords or any car spares shop (gives two coats) is an easy solution but do remember to carefully mask off the contacts rail under the top bar.

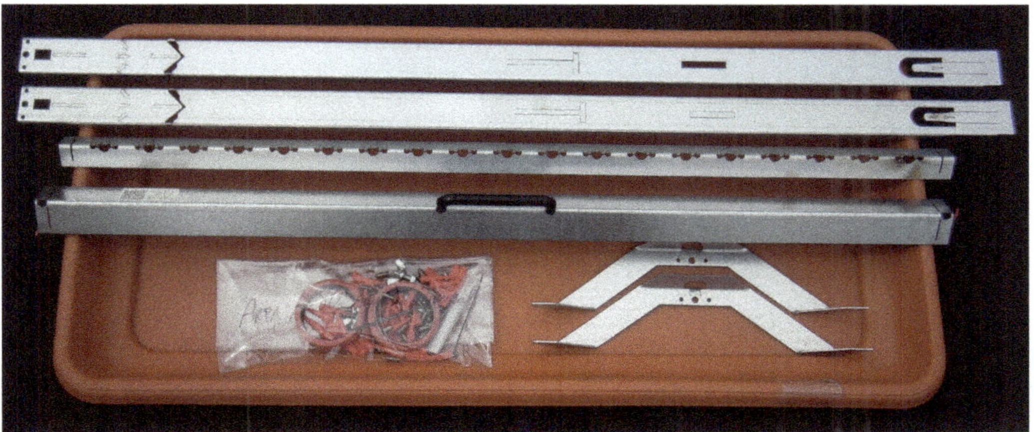

Figure 2 - A SANVE *Arpa* out of the box… Photograph – A Durham
These are well engineered units that have withstood the vagaries of the English weather in my Cambridgeshire 'test' apiary over two summers.

When the paint is dry and fully hardened, the parts are mostly push-fit. That was the easy bit. It won't be fully held together until the wires are strung. You must watch the SANVE video on assembly and wiring. Follow the wiring method, copying the demonstrator's technique exactly, and you soon get the hang of stringing the 40-odd wires and springs.

Go onto the SANVE website or search in YouTube: Montaje *arpa* universal y agua A-60 y A-80, SANVE, Unboxing and assembly - https://www.youtube.com/watch?v=oV_VEwiF1kc

Some essential tips on assembly are:

- ▸ Keep the fiddly parts (springs and tiny steel balls) over a tray with a towel on it while you thread the wire. It is very easy to drop the steel balls and they readily bounce and disappear for ever.

▸ Keep a tight control on the wire as you unreel it, do not let it come off the reel into loops or you will never unravel the consequent coils without kinking the wire. Keep an elastic band on the wire reel to stop that happening when you put it down.

▸ If you have bought a USB connection you cannot go wrong but the 12-volt option has two crocodile clips to attach to a car battery's terminals.

N.B. MAKE SURE YOU OBSERVE THE CORRECT POLARITY OR YOU CAN DESTROY THE POWER UNIT.

12.6.12 - Buy in the UK or E.U.?

Let's assume you decide to buy a harp either fully assembled or flat-pack (i.e., assemble and string the wires yourself). The first decision is whether to buy in the UK or buy in the E.U. and import it.

If you regularly travel to Europe by car, the obvious choice would be to buy over there and bring the *harpes/arpas* back as a personal import. ICKO Apiculture has stores off the two main routes into France from Calais. One at Totes near Rouen (215km from Calais just off A29) and another at Avelin near Lille (122km from Calais just off A1). There's bound to be an ICKO not too far away wherever you are staying in France.

In November 2024, most travellers can bring up to £390 worth of goods into the UK for personal use without paying UK VAT, duty and/or import tax. Each person has their own import allowance but the items bought back have to be obviously divisible, you can't bring back a single item costing £500 and claim 50% of the cost for one person and 50% for another. If you exceed the allowance you must declare them to Customs and pay the UK VAT (20%) plus any Customs duty (usually 2.5% on most items but in 2023 the *arpa* was zero-rated).

What is the situation if you decide to order online and have it/them shipped across? Let's say you wanted to buy two SANVE/Tartago *arpas* at 199 euros ea. plus shipping. On top of that cost there will be any UK import duty (2.5% but there was no import duty on an *arpa* in 2023) and you will have to pay VAT (20%) on the cost of the *arpa's* <u>plus</u> the shipping costs.

The shipping cost could be as high as 180 euros. The problem is that the harps are too bulky to go in a parcel via Colissimo and therefore have to go on a pallet. A possibility for a bulk order perhaps but not for an individual.

Then you must decide whether to pay the import duty etc. direct to HMRC (customs) or have the shipper act on your behalf. It's much simpler to let the shipping company handle this for you and there will be a fixed fee for that service.

It's actually a lot simpler than it sounds but obviously the much cheaper Spanish or French harp has gone up by 50% with all the extras. As the shipping tends to be per consignment up to a weight limit and the shipper's import fee is usually flat rate, the real extra cost on a bulk order is the UK VAT and any customs import duty.

12.6.13 – A Last Word about Electric Harps

As with any other system with multiple components, get it tested early in the season so you don't discover something doesn't work when you need it.

'Electric harps contribute significantly to mitigate the impact of this invasive species on apiaries… deployed in tandem with additional measures such as access to food sources during periods of highest predation'

Journal of Pest Management Science 18 August 2022

Wileyonlinelibrary DOI 10.1002/ps.7132

Research Article - Effectiveness of electric harps in reducing Vespa Velutina predation pressure and consequences for honey bee colony development

Authors: Sandra Rojas-Nossa, Damian Dasilva-Martins, Salustiano Mato, Carolina Bartolome, Xulio Maside and Josefina Garrido

12.7 - Defending the Hive

12.7.1 - The Entrance Restrictor

Some beekeepers are surprised when they leave their bees in the latter half of October, only to find colonies or even whole apiaries wiped out by the hornet when they return to them. *Muselieres* may have been removed prematurely but even if they have been left in place, if the hornet can get into the hive when the bees are clustered and there is enough of them to mount an attack, then the hive is lost (see inset below, where I have changed her name).

Le Parisien - December 2020:

'Three quarters of her hives are lost. At the beginning of November, in the space of two weeks, an attack by Asian hornets decimated the bee colonies installed by Claudette LaFranc in her village in the Eure.

Still shocked by her loss, Claudette had never experienced such a situation at this time of year but autumn broke heat records…

She had removed her traps and the brambles she normally puts around the hive eantrances…

The mistake that has been made is that the beekeeper has overlooked that the hornet can still be very active in late October and well into November. That period may even be extended if there has been a warm and extended autumn (often referred to as an 'Indian Summer').

In addition, by September hornet numbers are increasing and the *museliere* becomes useless because the hornets can get through larger size mesh and if they cannot get through because the mesh is too small they camp out on the mesh itself. I have counted over 40 hornets on a photograph of one *museliere*.

You must prevent the hornets entering the hive entrance en masse. The way to do that is to use the NICOT-type 5.5mm entrance restrictor, it costs just a couple of pounds from E.H.Thorne (search E H Thorne /Pests / Asian Hornet / MP floor entrance fittings). Make sure you pin it in place.

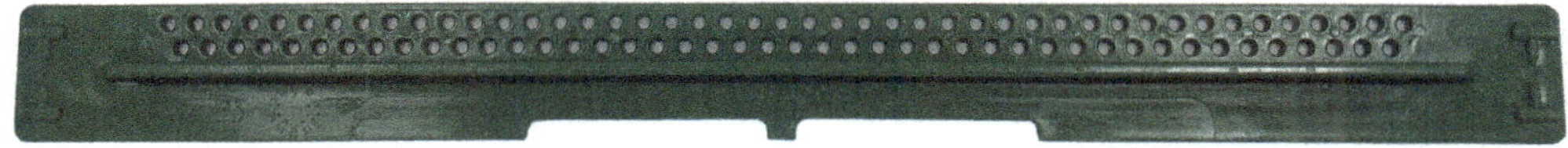

E H Thorne Green hornet entrance restrictor (£1.80 – 2024) Photo supplied by E H Thorne

Do not confuse the 5.5mm green hornet guard with the NICOT White guard which has 8.5mm openings. Its entrances are too big and the hornets just go through. Furthermore, some metal guards look like hornet guards but the gaps are 6.5mm and they are mouse guards. Too many beekeepers confuse hornet guards with mouse guards, a very common mistake.

EH Thorne – MP Floor with White Guard – Photo supplied by E H Thorne

The NICOT guard is not designed for the standard National Hive and unless you want to buy new floors, you need to be very careful if fitting it to a national hive. It is essential to check the gap is correct or you can trap your bees in the hive, I speak from experience! Also remember that the 5.5mm guard will trap drones in the hive. Don't leave them on over the whole winter – swap to mouse-guards. The bees need to be able to clean the hive.

If you are really strapped for cash, you could just use a strip of insert board but you must set the gap correctly with a 5.5mm drill (a standard drill size).

<u>This is very important</u>. There is a lot of misunderstanding about the entrance restrictor. Remember it is a last-ditch defence, the hornet guard does nothing to stop predation losses outside the hive nor does it deal with the hidden damage caused by stress. Its sole purpose is to prevent the hornet getting into the hive. It restricts foraging and it can trap drones in the hive if put on at the wrong time (but you may just have to sacrifice them to save the colony).

12.7.2 - The Veto-Pharma ApiShield

One of the author's four Veto-Pharma ApiShield traps. Photograph by the author.

The ApiShield works on the principle that the under-floor entrance, replacing the OMF, provides a more defensible entrance. The 4 round entrances on each side have 7mm Canadian cone escapes to trap both wasps and Asian hornets in a trapping chamber (the four holes at the back are ventilation holes only and have mesh screens).

I speak from the experience of owning four of these. It is very good for trapping wasps/robber bees and in tests by INRA it trapped a good number of Asian hornets in the autumn. However, it has weaknesses:

> ▶ It is evident that the complicated design has forced economies in the quality of the wood. It must be treated and the thin plywood

has a habit of warping in our damp/wet climate. If you accidentally pull out the top screen with a hive on top, you'll never get it back in properly.

▶ The two sides are prone to warp out at the back and need to be reinforced.

▶ The cones escapes have a nasty habit of getting clogged with spider webs and need cleaning out regularly.

▶ If the top sliding screen around the tunnel entrance warps even slightly it opens up a hiding place for wax moth.

▶ Perhaps the biggest weakness is that the hornets are trapped immediately under the brood box and that stresses the bees.

▶ It's expensive and in my view, the money is better spent on the *harpe electrique*.

Canadian cones escapes need regular cleaning of spider webs

13 - Trojan Bait Trapping

13.0 - Introduction

I have left this subject to last because whilst it is the most obvious solution to the Asian hornet, it is by far the most controversial and has a very real risk not only to the environment but also to beekeeping itself.

First, what is trojan bait trapping? It is where you catch hornets, dose them with an insecticide that the hornets take back to the nest, which is then poisoned. If you cannot find the nests to poison them, let the hornets do it for you! It has been used to deal with invasive wasps in a sensitive environment and I have watched a video of wildlife rangers putting out trays of minced meat dosed with Fipronil in order to do so. But if you bait food pellets you can end up killing the brood and not the queen or workers. The aim is to kill the whole nest.

A recent scientific paper (no. 69 in Further Reading) established that a dose rate far lower than that used in dog and cat flea treatment was effective in killing the brood in an Asian hornet nest. For some reason the researchers had dosed pellets of food for the workers to take back and feed to the brood, so the treatment is transitory, but I know of a beekeeper in France that sprays hornets captured in a net with a mixture of milk and fipronil, which in a mutual grooming, kills the workers and queen. Relief in the apiary lasts until hornets from other nests find it.

There is research going on in Spain and Italy to find a trojan bait treatment that can be approved. It's not straightforward as the poison has to be powerful enough to kill the nests but not so powerful or immediate that the trojan-bait-carrier doesn't make it to the nest.

The biggest problem is the poison itself. Fipronil is poisonous to birds and mammals that may feast on the dead nest and it degrades over time into something even more potent. Poison such as Sumithrin, a D-phenothrin, is less poisonous to mammals and degrades quickly but is very poisonous in the aquatic environment.

When but only when a suitable product has approval for trojan bait trapping, the rest is easy. Dry electric harps can capture large numbers of live hornets in the apiary. After a check for bycatch (absolutely essential) it would be easy to dose and release the hornets at dusk when they would return to their nests.

13.1 - If it works – Why Not Use Trojan baits?

As the law stands in 2024, it is illegal in the UK to release an invasive species, so once caught, the hornet may not be set free. Use of a veterinary medicine for other than its approved use will also land the misuser in trouble. Are these sufficient deterrents to a beekeeper trying to protect their colonies? I think not and the chances of being caught must be very slim. The beekeepers' argument in France is that millions of people dose dogs and cats that then urinate the medicine into the environment. Farmers pour pesticides by the tonne into the soil and whole hypermarket shelves are devoted to poisons to be used in the home or garden. If it works, why not?

If this guide has shown you anything, surely it is that whilst the pest is a serious problem that is not to be ignored, colony losses attributed to the hornet are often conflated with other factors that are nothing to do with the hornet and that it is often the beekeeper's lack of understanding or failure to act that has made matters worse for their bees. Fix that first.

Until we get levels of infestation that demand such a measure, the evidence is clear. This is a pest that can be managed with simple and relatively cheap measures combined with good beekeeping. It's like using a nuclear bomb as a first response. It doesn't make sense to do so. The fallout would be very damaging.

But if you are not convinced and that syringe of flea treatment is just too tempting, consider this. Beekeepers enjoy a unique position in the public perception. If the environmentalists targeting beekeeping because of its perceived threat to wild pollinators can accuse beekeepers of endangering biodiversity by the reckless use of insecticides, our unique position will be undermined.

It would certainly be an early Christmas for pesticide manufacturers who could point to our own use of them. Pesticides kill more honey bees than the Asian hornet. Any success with trojan bait trapping would be a pyrrhic victory. Furthermore, it would open the door to tighter regulation and put us on the wrong side of many research funding bodies.

14 – 'Blue sky' Thinking

Remember that the ideas behind the core defences detailed in this book came from beekeepers. I do not believe we need yet more reinvention of the same thing, what we need is for beekeepers to indulge in some 'blue sky' thinking and come up with something new.

Eugenio Pichel and the team at SANVE noticed that the hornet would instinctively attack an insect-sized dot or ball, mistaking it for a bee and that led to the Diana trap and to the *Arpa* CEBO. It is not just an improved *arpa*, it is a new idea combined with an existing one.

Earlier in the book I mentioned that Heather Mattila had found *Apis cerana* using dung around the entrance to its hive in order to deter *V.soror*.

We know the hornet follows olfactory clues and a particular favourite is the bee's Nasonov gland secretion, with which the bee marks the entrance to the hive. We are told that some natural recipes keep hornets away. I recall using oil of peppermint to mask whatever it was that was attracting robber bees to the top of one of my hives; it worked perfectly.

So, for my 'blue sky' idea, I shall try peppermint oil or lemongrass oil on a pad and regularly wipe it around the hive entrance. Nothing ventured, nothing gained.

The AAVO recommended a screen to keep the hornet away from the brood box floor. I instinctively went for a second lower screen as recommended by the AAVO but what if it were solid, what if the hive on its stand were a sealed unit? Could you stop any varroa climbing back up to the brood box? According to NOD, the manufacturers of MAQ's, the OMF doesn't do anything for ventilation. Would a sealed unit reduce the hive's olfactory signature?

Over to you.

15 - Overall Summary – Defences Against the Hornet

15.0 – Introduction

A short refresher on defences against the hornet with some reminder graphics but it's not a substitute for reading the detail in the main sections.

15.1 – Pre- Establishment.

Setting up an apiary? Think about layout, types of honey bee, get an idea for the scale of the problem you will face if the hornet gets established. Think about where you might place traps in the vicinity of your apiary (see Sections 3 &11).

15.2 – Hornet in Your Region But Not Yet in Your Area.

Make time to get educated about the hornet. The beekeeper who waits until the hornet is predating on their hives before acting, risks losing their bees. The beekeeper who doesn't understand what is happening, doesn't know what to do, will be amongst those who have the first tranche of colony losses (see Sections 7 – 9).

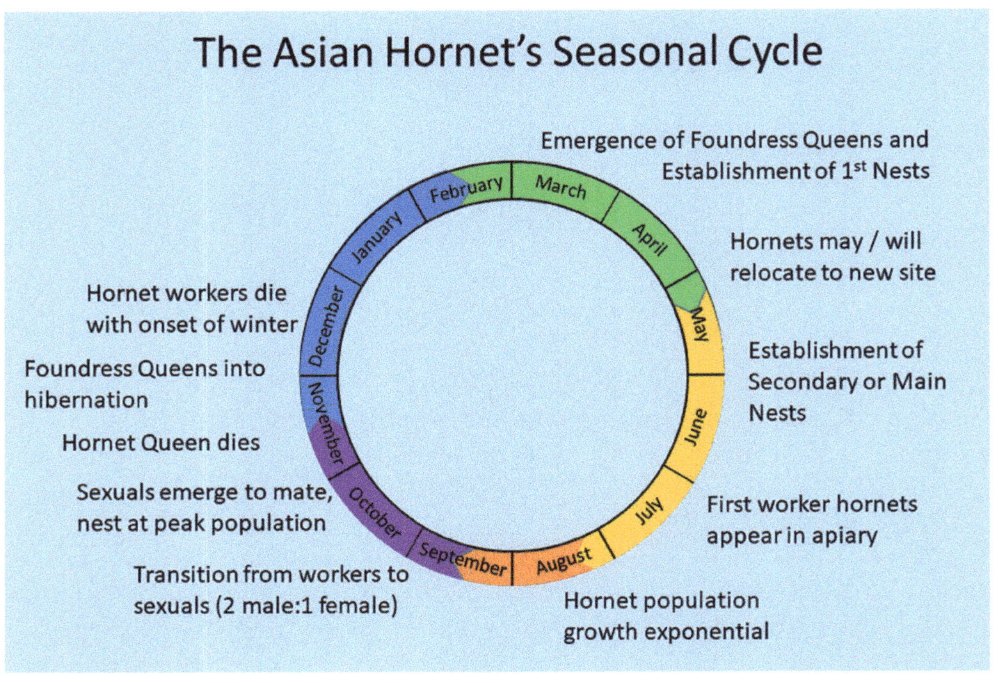

The Asian Hornet's Seasonal Cycle
January
February
March
April
May
June
July
August
September
October
November
December
Emergence of Foundress Queens and Establishment of 1st Nests
Hornets may / will relocate to new site
Establishment of Secondary or Main Nests
First worker hornets appear in apiary
Hornet population growth exponential
Transition from workers to sexuals (2 male:1 female)
Sexuals emerge to mate, nest at peak population
Hornet Queen dies
Foundress Queens into hibernation
Hornet workers die with onset of winter

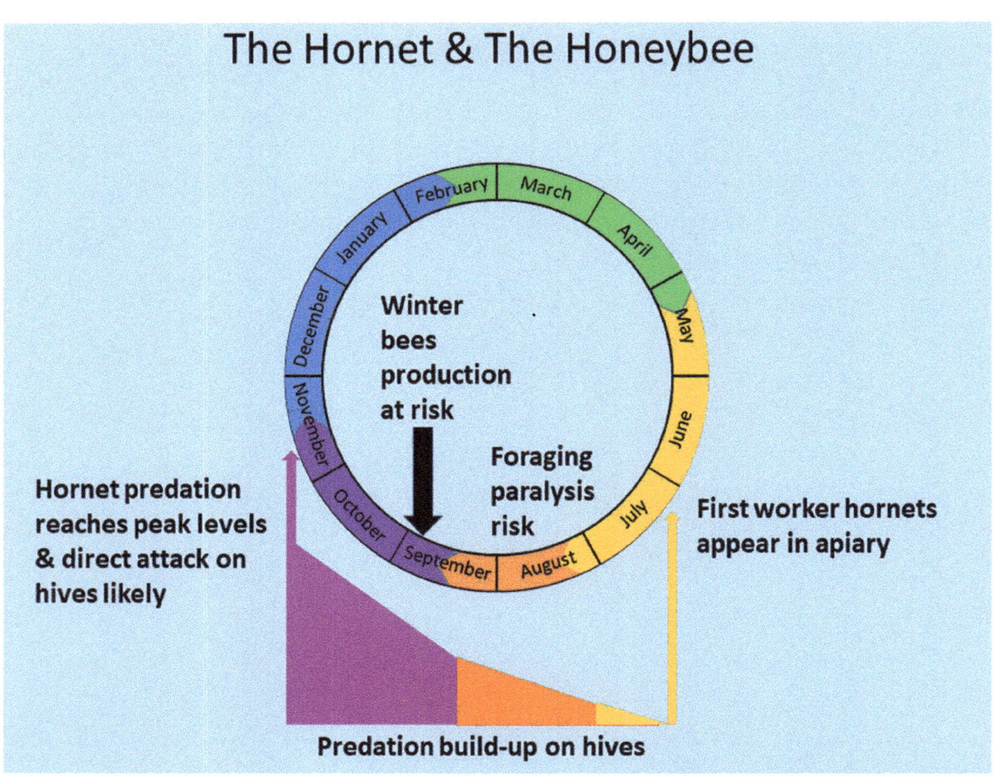

The Hornet & The Honeybee
January
February
March
April
May
June
July
August
September
October
November
December
Winter bees production at risk
Foraging paralysis risk
First worker hornets appear in apiary
Hornet predation reaches peak levels & direct attack on hives likely
Predation build-up on hives

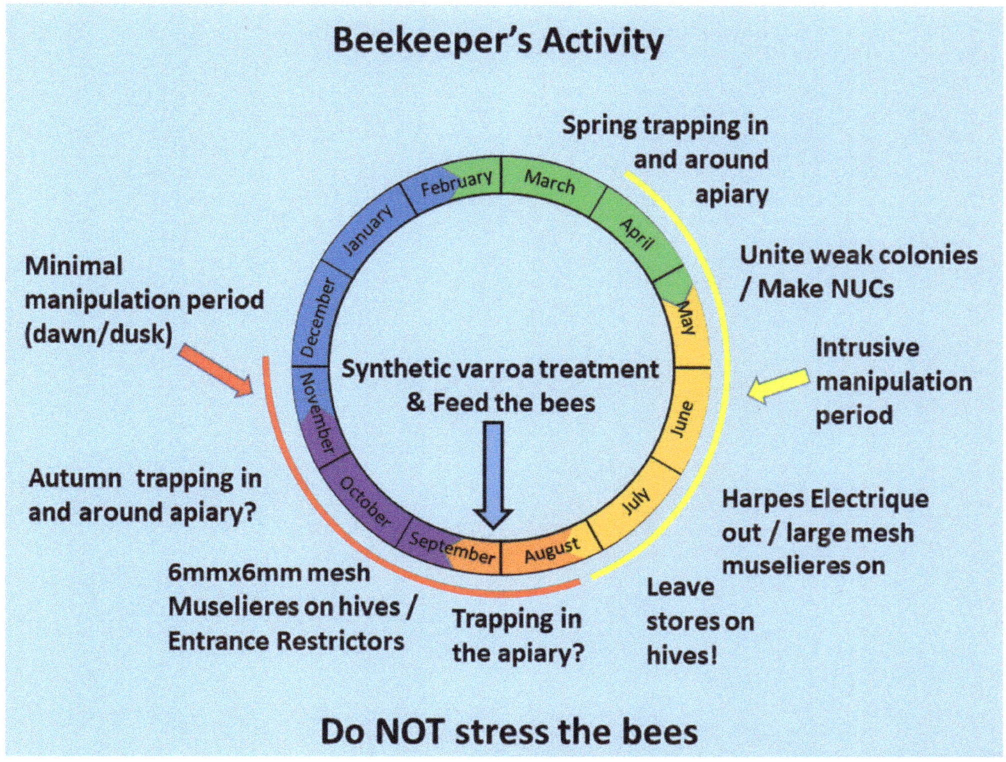

15.3 – Hornet in Your Area.

In the graphics below I have used traffic light colours – red is worse/bad, green is better/good and orange can go either way. Listed on the left are factors that will determine, for better or worse, the degree of predation that you may experience in your apiary (see Section 3.4). Some are down to location but the biggest factor in any year is the weather (see Section 7.2). Along the bottom of the graphic are the factors that affect the bees for better or worse and one size does not fit all; each colony will be different (see Section 9.3).

The arrival of the hornet in their area will have prompted the beekeeper to carry out selective spring trapping in the vicinity of the apiary (see Section 11.1).

The beekeeper who is prepared will have thought about their apiary's layout long before the hornet arrives and will have implemented some measures in advance or be ready to implement others (e.g., screening) (see Section12.2)

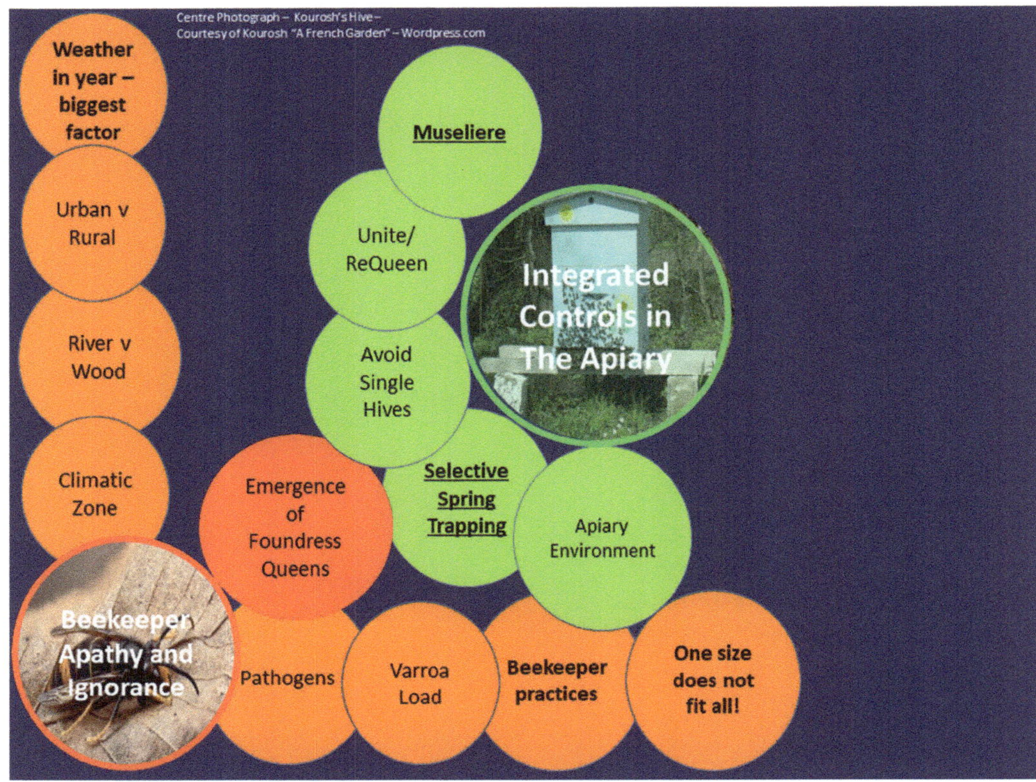

Summary of factors prior to start of predation in apiary

15.4 Predation on Hives

Once predation starts, the beekeeper will observe their hives closely. A low level of predation will prompt the installation of muselieres, with entrance restrictors ready to go on in the autumn as soon as the drones are out or when clustering starts and no guard bees are at the entrance. Stores will have been left on the hive after the main honey crop is taken off and the beekeeper will be ready to feed the bees (see Sections 9.3, 12).

The beekeeper who has invested in the *harpe electrique* may decide to deploy them early and even as an alternative to the museliere but be guided by what is happening at the hive entrance. Don't overlook the risks to swarming bees. Use the French layout for your harps. You may find that you don't need to deploy wet bait traps at all (see Section 12.5).

Summary of factors after start of predation in apiary

In the apiary itself we have two scenarios: Low/medium level predation when simple measures will likely suffice to meet the primary objective.

Get the *muselieres* on as soon as you see hawking near the hives but don't put out wet-bait traps unless it is necessary as you could attract even more hornets to the apiary. Don't forget to screen under the OMF's and don't forget the entrance restrictor (hornet 5.5mm) when you get into September.

You bees should always have plenty of stores; we are trying to avoid the bees getting it into their minds that food income is a problem because they will get stressed and may stop the queen egg-laying or worse.

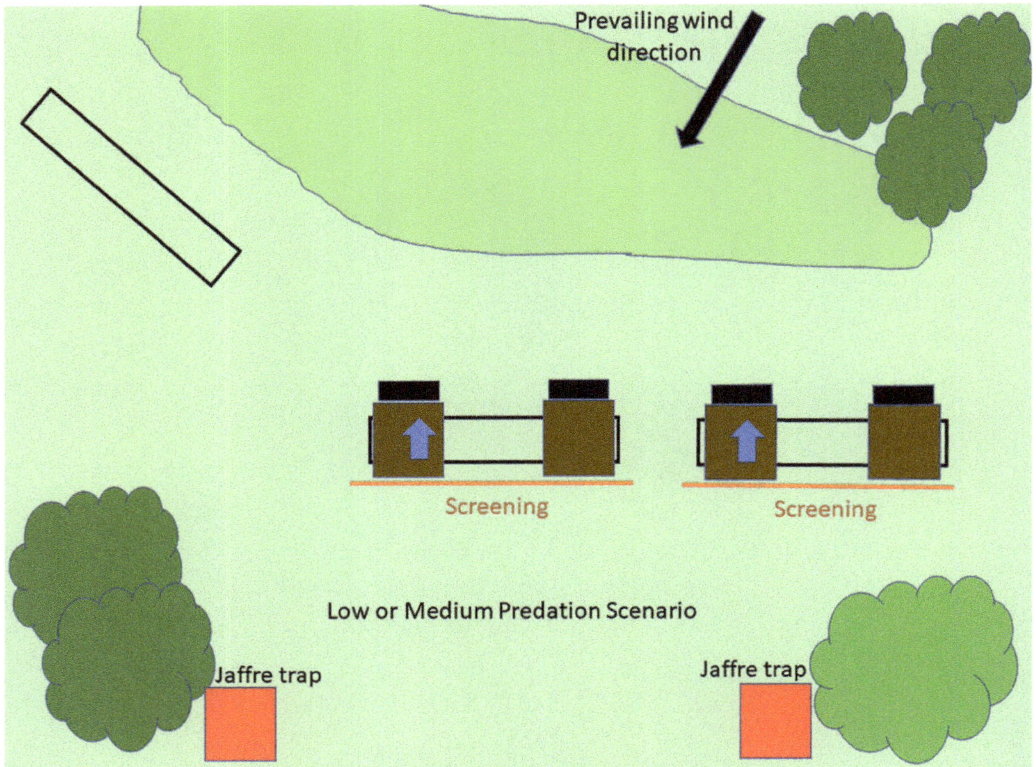

Defences deployed in Low-Medium Predation Scenario

If predation increases to a high level (see graphic below) the beekeeper must consider employing additional measures, deploying large capacity traps and the *harpe electrique*. Putting a screen behind hives helps to remove ambush hiding- spaces and a screen in front channels the hornet towards harps.

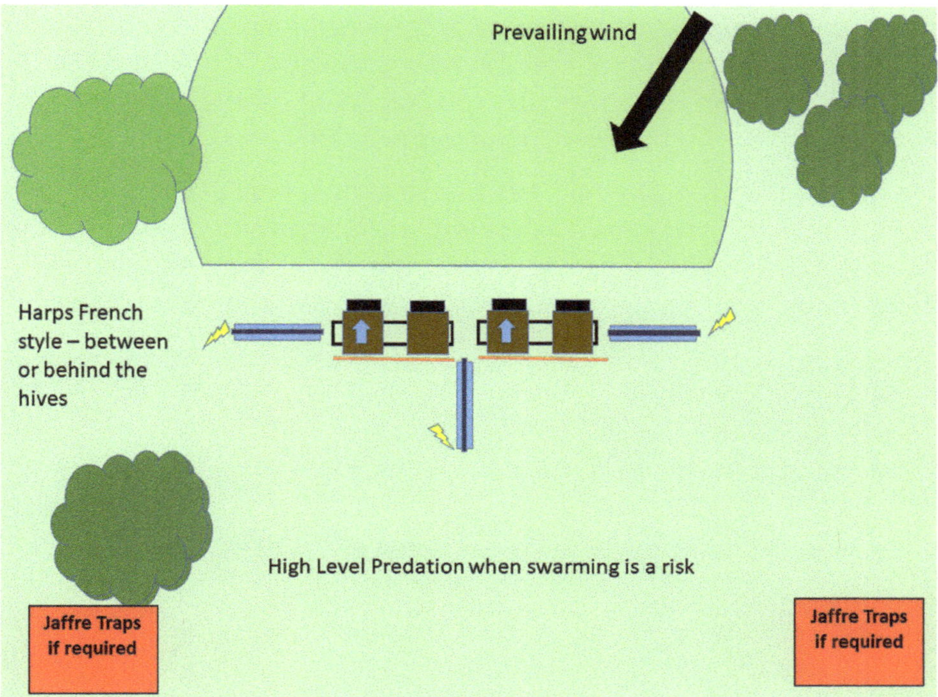

Defences deployed in when swarming is a risk

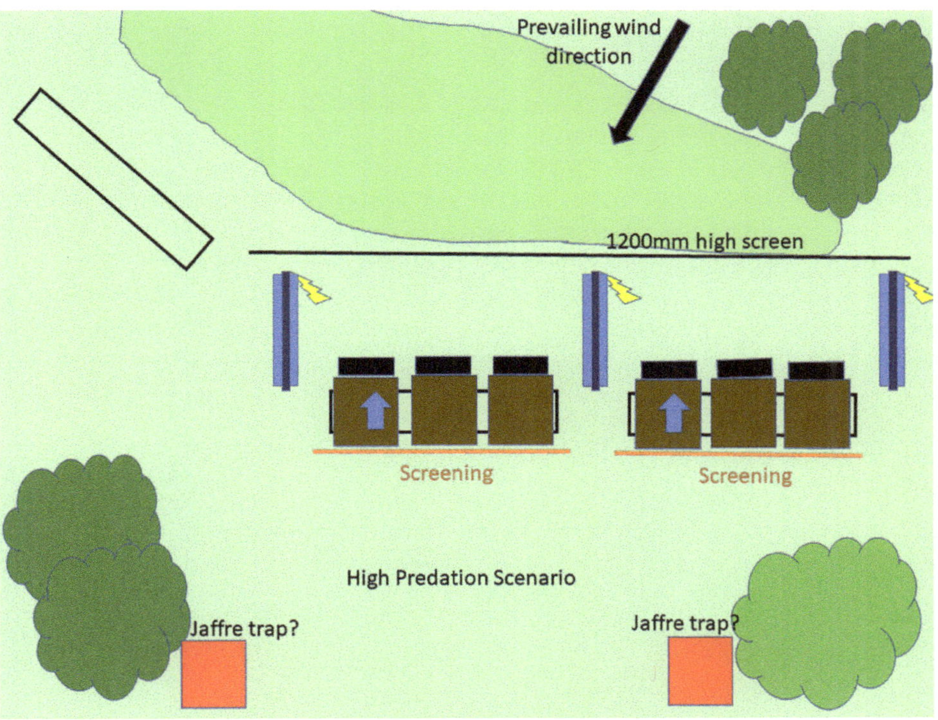

Defences deployed 'Spanish style' in High Predation Scenario

It's not a good idea leave everything out over winter, nor is it a good idea to wait until summer to find that new *museliere* doesn't fit the way you expected it to or that an HV power unit has burnt out and you didn't remember that when you put everything in the shed.

I think it's a good idea to train the bees to use a museliere prior to the hornet's appearance and I have no problem with harps and bees but not when there is a risk of swarming. So, at the same time that I am getting spare hives ready for any swarms, I get everything out and have a 'dry run'.

The author's apiary in March. The solar panel & central power supply can be seen to the right with its battery housed in a spare NUC. The harps are in the Spanish position to check cable runs but they will be moved back and at this stage there is no need to fill water baths. Muselieres are on for demonstration only – it's too early and they could interfere with swarming and foraging. The screen in front of the hives was new for 2025 and will hopefully screen our nearby raised vegetable beds from over-enthusiastic young guard-bees! The cloth screens on top of the hives (introduced in 2018) double-up as sunshades or to screen the rear of the hives as required.

Acknowledgements

I could not have produced this guide without the following individuals and organisations making information, reports and papers available.

Asian Hornet Beekeepers Guide
Acknowledgements

Museum National de la Histoire Naturelle

Institut National de Recherché pour l'Agriculture, l'alimentation et l'Environment

Fédération Départementale des Groupements de Défense contre les Organismes Nuisibles

Fédération Nationale des Organisations Sanitaires Apicoles Départementales

Union Nationale de l'Apiculture Français

Syndicat National D'Apiculture

Universitiy of Vigo

(and many other French Beekeeping and Research Organisations/Associations)

Eugenio Pichel of SANVE (España), Frederique Ripet, Dennis Jaffre, Kouroush

Andre Lavingotte, Eric Le Bervet, Randy Oliver (USA), Dr. Sandra Rojas-Nossa

Les Apiculteurs Français

I am very grateful to Ann Chilcott for proof-reading the book. Any errors are mine and most likely a last-minute tweak that did not go past her eagle eye.

I am not a good photographer and so I have licensed some photographs from Adobe stock.com which allows me to use them in a textbook, as long as the primary value of the product is not the photograph itself, and the product is not reproduced more than 500,000 times. The cover photograph required an extended licence.

If I have one last entreaty to make, it is that the British beekeeper does not waste time reinventing a wheel that has already been invented by our fellow beekeepers in France and Spain.

Further Reading and References

Section 1 – Introduction - Further Reading:

1.   **Vespa velutina: traits and impacts of a successful invasive alien species** – S. Rojas-Nossa, S.Mato, J. Garrido - Article *in* Ecosistemas · August 2021 - DOI: 10.7818/ECOS.2159 (In Spanish)

2.   **The invasive hornet Vespa velutina affects pollination of a wild plant through changes in abundance and behaviour of floral visitors** – Sandra Victoria Rojas-Nossa, Maria Cavivino-Cancela - Article *in* Biological Invasions · August 2020 DOI: 10.1007/s10530-020-02275-9

3.   **Vespa velutina: a new invasive predator of honey bees in Europe** - Karine Monceau • Olivier Bonnard • Denis Thiery 2013 - Article *in* Journal of Pest Science · March 2014 - DOI: 10.1007/s10340-013-0537-3

Section 2 – Background and Sources - Further Reading

4.   **Vespa velutina: An Alien Driver of Honey Bee Colony Losses -** Article in Diversity · January 2020   DOI: 10.3390/d12010005

5.   **Frelon Asiatique 12 ans après son installation, état des lieux et méthodes de lutte -** UNAF – Abeilles et Fleurs – Mars 2017 Hors Serie Special -

6.   **Guide Practique – FRELON ASIATIQUE COMMENT AGIR À L'ÉCHELLE DE MA COLLECTIVITÉ ? –** https://www.unaf-apiculture.info/IMG/pdf/guidefrelon.pdf

Section 3 – Thinking Well Ahead of the Hornet's Arrival - Further Reading:

7.   **Honey bees (*Apis cerana*) use animal feces as a tool to defend colonies against group attack by giant hornets - (*Vespa soror*).** Mattila HR, Otis GW, Nguyen LTP, Pham HD, Knight OM, Phan NT (2020) PLoS ONE 15(12): e0242668. https://doi.org/10.1371/journal.pone.0242668

8.   **Predicting the invasion risk by the alien bee-hawking Yellow-legged hornet Vespa velutina nigrithorax across Europe and other continents with niche models** - Claire Villemant, Morgane Barbet-Massin, Adrien Perrard, Franck Muller, Olivier Gargominy, Frédéric Jiguet, Quentin Rome - Biological Conservation 144 (2011) 2142–2150

9. **Predicting species distribution combing Multi-scale drivers** – Fournier A, Barbett-Massin M, Rome Q – Global Ecolgy and Conservation 2017 – DOI:10.1016/j.gecco.2017.11.002

10. **Can species distribution models really predict the expansion of invasive species** – Barbet-Massin M, Rome Q, Villemant C, Courchamp F, - PLOS One 2018 - DOI.org.10.1371/journal.pone0193085

11. **Climatic Niche Differentiation between the Invasive Hornet Vespa velutina nigrithorax and Two Native Hornets in Europe, Vespa crabro and Vespa orientalis** - Simone Lioy, Luca Carisio, Aulo Manino and Marco Porporato – Diversity **2023**, 15, 495. https://doi.org/10.3390/d15040495

12. **Rapid spread of the invasive yellow-legged hornet in France: the role of human-mediated dispersal and the effects of control measures** - Christelle ROBINET, Christelle SUPPO, Eric DARROUZET - Journal of Applied Ecology · June 2016 - doi: 10.1111/1365-2664.12724

13. **Monitoring & Control Modalities of a honey bee predator, the yellow-legged hornet Vespa Velutina nigrithorax (Hymenoptera:Vespidae)** Rome Q, Perrad A, Muller F, Villemant C – Aliens 31/2011

Section 4 – Scientists and Solutions · Further Reading:

14. **The Asian Yellow-legged Hornet: The implacable advance of a bee-killer** by Karine Monceau & Denis Thiery - Article *in* British Wildlife · December 2017

15. **Production of Early Diploid Males by European Colonies of the Invasive Hornet Vespa velutina nigrithorax.** Darrouzet E, Gévar J, Guignard Q, Aron S (2015) PLoS ONE 10(9): e0136680. doi:10.1371/journal.pone.0136680

16. **Triploid females and diploid males: underreported phenomena in *Polistes* wasps?** A.E. Liebert, R.N. Johnson, G.T. Switz and P.T. Starks – Insects Sociaux - Insect. Soc. 51 (2004) 205–211 0020-1812/04/030205-07 DOI 10.1007/s00040-004-0754-0 - © Birkhäuser Verlag, Basel, 2004

17. **Viruses in the Invasive Hornet Vespa velutina** Anne Dalmon, Philippe Gayral, Damien Decante, Christophe Klopp, Diane Bigot, Maxime Thomasson, Elisabeth A Herniou, Cédric Alaux and Yves Le Conte – Article in Viruses **2019**, 11, 1041; doi:10.3390/v11111041

18. **Assessment of the consumption of the exotic Asian Hornet Vespa velutina by the European HoneyBuzzard Pernis apivorus in southwestern Europe** – Rebollo S, Diaz-Aranda LM, Avila JAM, Hernandez Garcia M, et al (7 authors)   August 2023 - Bird Study 70(3):1-15 DOI:10.1080/00063657.2023.2244258

19. **Can parasites halt the invader? Mermithid nematodes parasitizing the yellow-legged Asian hornet in France** - Claire Villemant, Dario Zuccon, Quentin Rome, Franck Muller, George O. Poinar Jr and Jean-Lou Justine – 2015 - PeerJ 3:e947; DOI 10.7717/peerj.947

20. **Designing a sex pheromone blend for attracting the yellow-legged hornet *Vespa velutina*), a pest in its native and invasive ranges worldwide -**
Ya-Nan Cheng, Ping Wen, Ken Tan, and Eric Darrouzet - Entomologia Generalis, Vol. XX (2022), Issue X, XXX–XXX PrePub-Article Published online February 2022 - DOI: 10.1127/entomologia/2022/1395

21. **A *Beauveria bassiana* strain naturally parasitizing the bee predator *Vespa velutina* in France -** Juliette Poidatz, Rodrigo Javier Lopez Plantey and Denis Thiéry - Entomologia Generalis, Vol. 39 (2019), Issue 2, 73–79
Published online 11 September 2019 - © 2019 The Authors www.schweizerbart.de - DOI: 10.1127/entomologia/2019/0690

22. **Predicting the spatio-temporal dynamics of biological invasions: Have rapid responses in Europe limited the spread of the yellow-legged hornet (*Vespa velutina nigrithorax*)? -** Richard M. J. Hassall | Bethan V. Purse| Louise Barwell | Olaf Booy |Simone Lioy | Stephanie Rorke | Kevin Smith4| Riccardo Scalera |Helen E. Roy - © 2024 The Author(s). *Journal of Applied Ecology* - DOI: 10.1111/1365-2664.14829

23. **Science communication is needed to inform risk perception and action of stakeholders** - March 2020 - Journal of Environmental Management 257(109983) DOI:10.1016/j.jenvman.2019.109983 Authors: Fabrice Requier, Alice Fournier, Quentin Rome, Eric Darrouzet

24. **Twenty years of attempting to control the Vespa velutina invasion: will we win the battle?** Thiery D, Monceau K - Article *in* Entomologia Generalis · July 2024 DOI: 10.1127/entomologia/2024/2731

25. **Economic costs of the invasive Yellow-legged hornet on honey bees** - Fabrice Requier, Alice Fournier, Sophie Pointeau, Quentin

Rome, Franck Courchamp - Science of the Total Environment journal 2023 - https://doi.org/10.1016/j.scitotenv.2023.165576

**Section 5 – Nest Destruction & Wide-area Spring Trapping -
A Job for Beekeepers?
Further Reading:**

26. **Searching for nests of the invasive Asian hornet (Vespa velutina) using radio-telemetry** - Peter J. Kennedy, Scott M. Ford, Juliette Poidatz, Denis Thiéry & Juliet L. Osborne - COMMUNICATIONS BIOLOGY | (2018) 1:88 | DOI: 10.1038/s42003-018-0092-9

27. **An assessment of the invasion of *Vespa velutina* in Flanders, northern Belgium -** Jasmijn Hillaert, Gemma Burbui, Sander Devisscher, Emilie Gelaude, SanneVan Donink & Tim Adriaens - Vespid task force COLOSS, 23-25 October 2023, Pisa

**Section 6 – The Hornet and Humans – The Risk to Human Health -
Further Reading:**

28. **Defensive behavior of the invasive alien hornet Vespa velutina nigrithorax against potential human aggressors** – Moon Bu Choi - Article *in* Entomological Research · March 2021 DOI: 10.1111/1748-5967.12514

29. **Defensive behaviour of the invasive alien hornet, Vespa velutina, against color, hair and auditory stimuli of potential aggressors** Moon Bo Choi, Eui Jeong Hong and Ohseok Kwon - PeerJ 9:e11249 DOI 10.7717/peerj.11249

30. **Human Fatalities Caused by Hornet, Wasp and Bee Stings in Spain: Epidemiology at State and Sub-State Level from 1999 to 2018 -** Xesús Feás - Biology **2021**, 10, 73. https://doi.org/10.3390/biology10020073

31. **Vespa Velutina Nigrithorax: a new causative agent for anaphylaxis** Ana Isabel Tabar, Sendy Chugo, Alejandro Joral, Maria Teresa Lizaso, Susana Lizarza, Maria Jose Alvarez-Puebla, Esozia Arroabarren, Catalina Vela, Manuel Lombardero - From Food Allergy and Anaphylaxis Meeting 2014 - Tabar et al. Clinical and Translational Allergy 2015, 5(Suppl 3):P43 - http://www.ctajournal.com/content/5/S3/P43

32. **Anaphylaxis to *Vespa velutina nigrithorax*: Pattern of Sensitization for an Emerging Problem in Western Countries** - Vidal C, Armisén M, Monsalve R, González-Vidal T, Lojo S, López-Freire S, Méndez P, Rodríguez V, Romero L, Galán A, González-Quintela A, - J Investig Allergol Clin Immunol 2021; Vol. 31(3): 228-235

doi: 10.18176/jiaci.0474

33. **Clinical Features of Severe Wasp Sting Patients with Dominantly Toxic Reaction: Analysis of 1091 Cases**. Xie C, Xu S, Ding F, Xie M, Lv J, et al. (2013) PLoS ONE 8(12): e83164. doi:10.1371/journal.pone.0083164

34. **Deciphering the Venomic Transcriptome of Killer-Wasp Vespa velutina** - Zhirui Liu1, Shuanggang Chen, You Zhou, Cuihong Xie, Bifeng Zhu, Huming Zhu, Shupeng Liu, Wei Wang, Hongzhuan Chen & Yonghua Ji – 2015 – Scientific Reports - DOI: 10.1038/srep09454

35. **The Asian wasp Vespa velutina nigrithorax: Entomological and allergological characteristics** - Carmen Vidal 2021 *Clinical & Experimental Allergy* published by John Wiley & Sons Ltd. DOI: 10.1111/cea.14063

36. **Medical consequences of the Asian black hornet (Vespa velutina) invasion in Southwestern France** - Luc de Haro, Magali Labadie, Pierre Chanseau, Claudine Cabot, Ingrid Blanc-Brisset, Françoise Penouil - National Coordination Committee for Toxicovigilance – 2009 Toxicon doi:10.1016/j.toxicon.2009.08.005

37. **Vespa velutina nigrithorax venom allergy – Real life clinical practice** - Joana Miranda MD , Ana M. Mesquita MD , Jos´e Pl´acido MD , Alice Coimbra MD , , *Annals of Allergy, Asthma Immunology* (2022), doi: https://doi.org/10.1016/j.anai.2022.07.020

38. **Allergy to stings and bites from rare or locally important arthropods: Worldwide distribution, available diagnostics and treatment –**
Gunter Johannes Sturm | Elisa Boni | Darío Antolín-Amérigo | Maria Beatrice Bilò | Christine Breynaert | Filippo Fassio | Kymble Spriggs |Arantza Vega | Luisa Ricciardi | Lisa Arzt-Gradwohl | Wolfgang Hemmer - © 2023 The Authors. *Allergy* published by European Academy of Allergy and Clinical Immunology and John Wiley & Sons Ltd.

39. **Ocular Lesions Other Than Stings Following Yellow-Legged Hornet (*Vespa velutina nigrithorax*) Projections, as Reported to French Poison Control Centers** - Hervé Laborde-Castérot,MD, PhD; Eric Darrouzet, PhD; Gaël Le Roux, PharmD; Magali Labadie, MD; Nicolas Delcourt, PharmD, PhD; Luc de Haro,MD, PhD; Dominique Vodovar, MD, PhD; Jérôme Langrand, MD; for the French Poison Control Centers Research Group – JAMA Ophthalmology | Brief Report 2020

40. **Epidemiology and risk factors of self-reported systemic allergic reactions to a Hymenoptera venom in beekeepers worldwide: a protocol for a systematic review of observational studies.** Carli T, Locatelli I, Košnik M, *et al. BMJ Open* 2022;12:e058130. doi:10.1136/bmjopen-2021-058130

41. **The prevalence of self-reported systemic allergic reaction to *Hymenoptera* venom in beekeepers worldwide: A systematic literature review and meta-analysis.** Carli T, Locatelli I, Košnik M, Kukec K. – Sciendo - Zdr Varst. 2024;63(3):152-159. doi: 10.2478/sjph-2024-0020. © National Institute of Public Health, Slovenia.

42. **Risk factors and indicators of severe systemic insect sting reactions**
Johanna Stoevesandt | Gunter J. Sturm | Patrizia Bonadonna | Joanna N.G. Oude Elberink | Axel Trautmann - *Allergy.* 2020;75:535–545. wileyonlinelibrary.com/journal/all © 2019 EAACI and John Wiley & Sons A/S. DOI: 10.1111/all.13945

Section – 7 The Asian Hornet – A Consummate Predator - Further Reading:

43. **Embryo, Relocation and Secondary Nests of the Invasive Species Vespa Velutina in Galicia (NW Spain)** – Animals MDPI (2022) doi.org/10.33390/ani12202781

44. **Vespa velutina Lepeletier, 1836 (Hymenoptera: Vespidae): first records in Iberian Peninsula** - S. Lo´pez, M. Gonza´ lez and A. Goldarazena - © 2011 The Authors. Journal compilation © 2011 OEPP/EPPO, Bulletin OEPP/EPPO Bulletin 41, 439–441

45. **Impacts of the invasive hornet *Vespa velutina* on native wasp species: a first effort to understand population-level effects in an invaded area of Europe** - Luca Carisio · Jacopo Cerri · Simone Lioy · Ettore Bianchi · Sandro Bertolino· Marco Porporato - Journal of Insect Conservation (2022) 26:663–671 https://doi.org/10.1007/s10841-022-00405-3

Section 8 – The Hornet and the Honey bee - Further Reading:

46. **Defensive behaviour of Apis mellifera against Vespa velutina in France: Testing whether European honey bees can develop an effective collective defence against a new predator** – Elsevier Behavioural Processes 2014 http://dx.doi.org/10.1016/j.beproc.2014.05.002

47. **Native Prey and Invasive Predator Patterns of Foraging Activity: The Case of the Yellow-Legged Hornet Predation at European Honey bee Hives** – Monceau K, Arca M, Lepretre L, Mougel F, Bonnard O, Silvain J-F, Maher N, Arnold G, Thiery D – PLOS One June 13, 2018 DOI.org/10.137/journal.pone.0066492

48. **Predation of the invasive Asian hornet affects foraging activity and survival probability of honey bees in Western Europe** - Fabrice Requier · Quentin Rome · Guillaume Chiron · Damien Decante · Solène Marion · Michel Menard · Franck Muller · Claire Villemant · Mickaël Henry - Journal of Pest Science 2018 https://doi.org/10.1007/s10340-018-1063-0

49. **Stress-Mediated Allee Effects Can Cause the Sudden Collapse of Honey Bee Colonies.** Booton, R.D., Iwasa, Y., Marshall, J.A.R. et al. (1 more author) (2017) Journal of Theoretical Biology. ISSN 0022-5193 https://doi.org/10.1016/j.jtbi.2017.03.009

50. **Timing of production of winter bees in honey bee (Apis mellifera) colonies** - H.R. Mattila, J.L. Harris and G.W. Otis 2001 Insects Sociaux Insectes soc. 48 (2001) 88–93 0020-1812/01/020088-06 $ 1.50+0.20/0

51. **Olfactory Attraction of the Hornet Vespa velutina to Honey bee Colony Odors and Pheromones**. Couto A, Monceau K, Bonnard O, Thiery, D, Sandoz J-C (2014)   PLoS ONE 9(12): e115943. doi:10.1371/journal.pone.0115943

52. **Predation pressure dynamics study of the recently introduced honey bee killer Vespa velutina: learning from the enemy**. Karine Monceau, Nevile Maher, Olivier Bonnard, Denis Thiéry. Apidologie, Springer Verlag, 2013, 44 (2), pp.209-221. 10.1007/s13592-012-0172-7. hal-01201288

53. **Density of predating Asian hornets at hives disturbs the 3D flight performance of honey bees and decreases predation success -** Juliette Poidatz | Guillaume Chiron | Peter Kennedy | Juliet Osborne | Fabrice Requier - © 2023 The Authors. *Ecology and Evolution* published by John Wiley & Sons Ltd. - https://doi.org/10.1002/ece3.9902

54. **Honey bee social collapse arising from hornet attacks -** Shihao Dong, Gaoying Gu, Jianjun Li, Zhengwei Wang, Ken Tan, Mingxian Yang, and James C. Nieh -

Entomologia Generalis Published online May 2023 - DOI: 0.1127/entomologia/2023/1825

55. **Monitoring Study in Honey bee Colonies Stressed by the Invasive Hornet Vespa velutina** - Ana Diéguez-Antón , María Shantal Rodríguez-Flores , Olga Escuredo and María Carmen Seijo - Vet.Sci. **2022**, 9, 183. https://doi.org/10.3390/vetsci9040183

Section 9 - The Asian Hornet – Honey bee Colony Mortality, Good Beekeeping and Healthy Colonies - Further Reading:

56. **Mortalités, effondrements et affaiblissements des colonies d'abeilles *(Weakening, collapse and mortality of bee colonies)*** AFSSA Report - April 2009 (French/English)
57. **Senate Information Report no.474 (2016-2017) by the Committee on Spatial Planning & Sustainable Development on the Commission into the fight against the decline of Pollinators** by Mr. Herve Maurey March 22nd 2017
58. **Vespa velutina: current situation and perspectives** – Karine Monceau, Denis Thiery - Conference Paper · November 2016 Researchgate

Section 10 – The Spring Trapping Debate - Further Reading:

59. **Performance of baited traps used as control tools for the invasive hornet Vespa velutina and their impact on nontarget insects** - Sandra V. ROJAS-NOSSA, Noelia NOVOA, Antonio SERRANO, María CALVIÑO-CANCELA - Apidologie © INRA, DIB and Springer-Verlag France SAS, part of Springer Nature, 2018 DOI: 10.1007/s13592-018-0612-0
60. **All That Glitters Is Not Gold: The Other Insects That Fall into the Asian Yellow-Legged Hornet Vespa velutina 'Specific' Traps**. Sánchez, O.; Arias, A. – Biology 2021, 10, 448. https://doi.org/10.3390/biology10050448
61. **Not just honey bees: predatory habits of *Vespa velutina* (Hymenoptera: Vespidae) in France** - Quentin Rome , Adrien Perrard , Franck Muller , Colin Fontaine , Adrien Quilès , Dario Zuccon & Claire Villemant (2021):Annales de la Société entomologique de France (N.S.), DOI: 10.1080/00379271.2020.1867005
62. **Chasing the queens of the alien predator of honey bees: A water drop in the invasiveness ocean** - Karine Monceau, Olivier Bonnard, Denis Thiéry – Open Journal of Ecology Vol.2, No.4, 183-191 (2012) http://dx.doi.org/10.4236/oje.2012.24022

63. **Evaluation of Asian Hornet (Vespa velutina) Trappability in Alto-Minho, Portugal: Commercial vs. Artisanal Equipment, Human Factors, Geography, Climatology, and Vegetation** – Mata F, Cano-Diaz c - Article *in* Applied Sciences · August 2024 DOI: 10.3390/app14177571

64. **Piégeage de printemps des fondatrices de frelon asiatique en Wallonie · Bilan de la campagne 2024 ·** Florian Bastin, Cyril Voss, Louis Hautier · Unité santé des Plantes et forêts – Centre wallon de Recherches agronomiques https://www.srawe.be/wp-content/uploads/2024/07/Campagne-de-piegeage-de-frelon-asiatique-wallonie-2024-vf.pdf

Section 12 – Defences in the Apiary · Further Reading:

65. **A biodiversity-friendly method to mitigate the invasive Asian hornet's impact on European honey bees** - Fabrice Requier · Quentin Rome · Claire Villemant · Mickaël Henry - Journal of Pest Science - https://doi.org/10.1007/s10340-019-01159-9

66. **Effectiveness of electric harps in reducing Vespa velutina predation pressure and consequences for honey bee colony development ·** Sandra V Rojas-Nossa, Damian Dasilva-Martins, Salustiano Mato, Carolina Bartolomé, Xulio Maside and Josefina Garrido – Pest Management Science – 2022 - (wileyonlinelibrary. com) DOI 10.1002/ps.7132

67. **Testing the selectiveness of electric harps: a mitigation method for reducing Asian hornet impact at beehives ·** CRISTIAN PÉREZ-GRANADOS, JOSEP M. BAS, JORDI ARTOLA, KILIAN SAMPOL, EMILI BASSOLS, NARCÍS VICENS, GERARD BOTA, NÚRIA ROURA-PASCUAL – Researchgate

68. ***Electrical* traps, so called harps, efficient and selective against *Vespa velutina* workers predating on hives ·** Denis Thiéry, Mónica Doblas-Bajo, Zoé Tourrain, Gaëtane Le Provost, Etelvina Núñez-Pérez - Article *in* Entomologia Generalis · October 2023 - DOI: 10.1127/entomologia/2023/2051

Section 13 – Trojan Bait Trapping · Further Reading:

69. **Efficacy of Protein Baits with Fipronil to Control Vespa velutina nigrithorax (Lepeletier, 1836) in Apiaries - Barandika, J.F.; de la Hera,O.; Fañanás, R.; Rivas, A.; Arroyo, E.; Alonso, R.M.; Alonso, M.L.; Galartza, E.; Cevidanes, A.; García-Pérez, A.L. – Animals 2023, 13, 2075. https://doi.org/10.3390/ani13132075**

Additional Further Reading:

70. An inhibitory signal associated with danger reduces honey bee dopamine levels - Shihao Dong, Gaoying Gu, Tao Lin, Ziqi Wang, Jianjun Li, Ken Tan,1 and James C. Nieh – Current Biology - https://doi.org/10.1016/j.cub.2023.03.072

Annex – Building a *Harpe electrique*

You can make your own *harpe* and designs are to be found on the internet. What follows is just my solution but it works and was tested for real when a surge of European hornet numbers in 2018 put my bees under severe pressure.

Home-made *harpe electrique* - Photograph: Andrew Durham

"the use of the muzzle and the electric harp is perfect in terms of efficiency and selectivity with a strong drop in stress on colonies"

National Association of Bee Health Groups' Report September 2023

Harps typically come in 60cm – 80cm widths. I recommend the 80cm width so that you don't leave a big gap between the harps and the fence in front of the hives.

There are a number of component parts to the *harpe electrique*, so we'll look at each part in turn.

Harpe electrique – Frame and Grid Wires

Overview

The *Harpe electrique*'s frame is the main part of the trap, housing the wires that will pass a high voltage shock to the hornet and causing it to fall into the water bath (wet) or trapping chamber (dry). Poor wiring can render the harp useless and it is very important to get that right.

The harp described is 80cm wide (internal dimension) and 100cm high (internal dimension). These dimensions are determined by:

> Width - 40 vertical wires spaced 2cm apart plus a 1cm gap at each side to allow for any protruding bolts/nuts. There is no point making the harp wider it just weakens the frame.

> Height – A standard pipe length is 2m. Stainless steel wire comes in reels 25m or 50m long. A 50m reels divided by 40 = 1.25m each wire, enough extra for knots and the return loop through a connector block plus 'tail to link to the next connector in that polarity.

Two triangles of plywood (painted) bolted to the frame will act as supports but make sure you leave space for the water trays underneath.

My advice is that few apiaries are completely level and a slope can mean that the depth at one end of the tray is insufficient, especially in the summer when the water will evaporate. I have found two baths to work better in such circumstances.

Component List/Budget

Qty 2 – 2m lengths 40mm dia. waste pipe B&Q £4 ea.
Qty 4 – 40mm 90 deg. bend fitting B&Q £2 ea.
Qty 2 – 1m M12 stainless steel threaded rod with qty 8 nuts and washers
       Toolstation or equiv. £10 ea.

Qty 4   – M8 L100mm bolts with nuts & washers B&Q £7 or use 38mm 'Terry clips'.

Qty 40 – circa 25mm - 40mm length expansion springs Amazon or Halfords £7.50

Qty 40 – 3amp connectors (12-way strips) Amazon £8 or solder-seal wire connectors.

Qty 40 – 140mm tie-wraps (pack of 50) B&Q £2.30

Qty 1   – 50m spool of stainless steel wire 316-grade circa 0.3mm dia. Crazy Wire Co. £7

Note: Do not buy coated wire or resistance wire e.g., NiChrome wire.

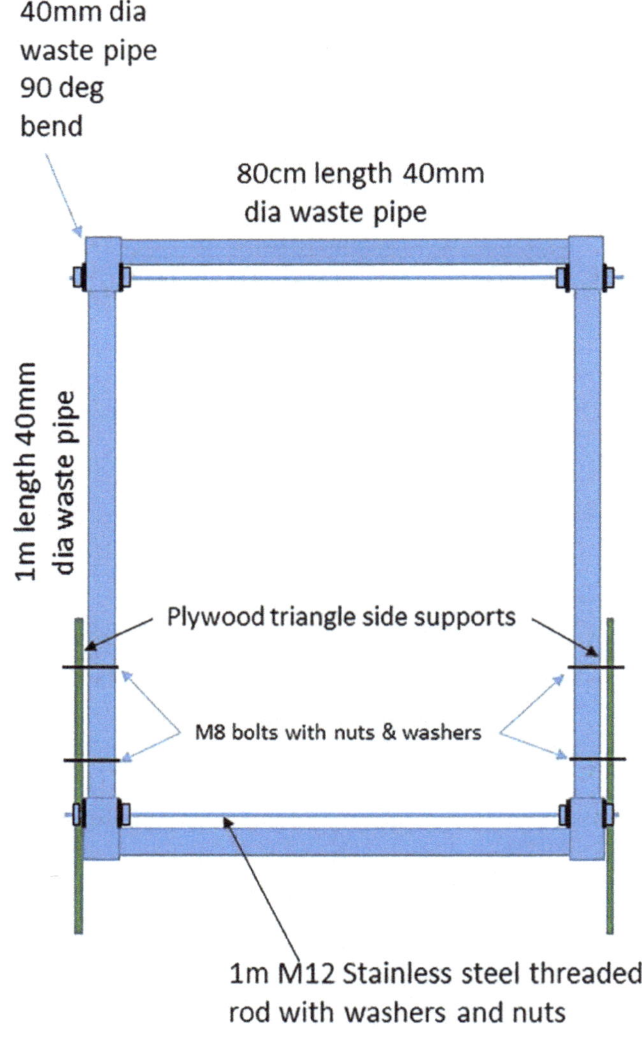

The harp's frame

Assembling the Frame - Tips

There is no need to solvent-weld the joints if you put the stainless steel rods through the 90-deg. bends and the mould 'seams' of the bends will help ensure you drill the holes in the right place.

Mark pipes before pushing them into the bends so you know they are fully home. If you don't get both sides the same the frame won't be square.

Use petroleum jelly to ease pipes into bends (or they can stick solid halfway in!).

Start by assembling the two vertical pipes and bends using a flat surface to ensure the bends on each end are accurately aligned with each other.

Drill the holes through the bends and pipes for the threaded rods.

Before inserting the top and bottom pipes, drill holes for HV supply wires and feed the wires through pipe.

Install the top and bottom pipes.

Install the threaded rods as shown in the diagram. You don't have to cut the rods if you are going to drill holes for them in the plywood supports. In fact, they can be useful to support the frame and in later versions I have used 38mm 'Terry Clips' instead of M8 bolts (makes it easier to take the plywood supports off and store the harp over winter).

If the rods extend beyond the frame don't forget to insulate them. They will be electrically live when the harp is operating.

When you have finished the harp, drill a couple of drain holes in the bottom bends to allow any water that gets in to escape.

Wiring the Frame

Some French designs use a continuous wire to avoid the poor-connections problem but, assuming an 80cm width and 20mm gap between wires, and each wire turning 180 degrees over the top and bottom insulated bars (you cannot loop over a bare metal rod - the wires will short out) by the time you have got to the end of the run there is so much friction between the wires and the bars it is impossible to get the tension right all the way across the harp.

It is better if each of the 40 wires is independently sprung.

Do not try to use zinc-plated or galvanised threaded rods as electrical conductors because they will corrode or you will get a bi-metal reaction with the stainless steel grid wires and end up with poor electrical joints. You can use stainless steel rods as conductors if you are using stainless steel wires.

A 'belt & braces' solution is to put a connector on each wire and link wires of the same polarity using jumper wires. It saves a lot of rewiring if the steel rods are not electrically good enough or become corroded over time.

The wires will be wired positive – negative- positive – negative etc. Each polarity's wire is going to be electrically insulated at one end by a tie-wrap loop, to which is attached the tension spring. The grid wire is attached to the other end of the spring (see diagram).

There is more than one way of doing the other end of the grid wire, the end that is going to be (electrically) connected to the stainless steel rod. You can either take the wire through a connector, loop it round the rod and bring it back through the connector OR you can take it through a connector (to connect a jump wire) and fix the wire to the rod. If you don't want a jump-wire, you can leave the connector off completely but then you are totally reliant on the wire-to-rod electrical connection.

In the diagram I have shown the wire looping round and coming back up because if you loop the wire round the rod rather than fix it, you can adjust that end of the wire along the rod as well as being able to move the tie-wrap loop at the other end. It's up to you.

If you use the connector to loop the wire round the rod and also use it for the jump-wire then you end up with all one polarity's connectors at the top and the other polarity's connectors at the bottom, as per diagram.

If you fix one end of the wire to the rod itself, you can choose to have all the jump wire connectors at the top under a rain screen.

When it comes to what connectors to use, the most obvious is a small 'choc-bloc' connector but there is the risk that they can become a moisture trap and thus a source of corrosion. A connector that I am trialling on one of my harps is the waterproof shrink tube that has internal seals and a low melting point solder. For now, we'll assume the use of the 'screw choc-bloc'.

0.5mm diameter stainless steel wire works but 0.3mm is better. It is thinner, less visible but it can be difficult to get the choc-bloc screws to grip the thinner wire.

The wire reel comes with the end of the wire sticking out through a slit in the reel and fixed under the label. Before you free that end, put a rubber band round the reel because what often happens is that the wire comes loose and unravels outside the reel. That is the perfect recipe for a loop in the wire that forms a kink and it is the bane of this wiring task. Always unreel wire with your finger carefully holding the wire tight to the rest of the wire on the reel. Do not let that pressure off and when you have the desired length of wire, slip the elastic band over the reel so that the wire cannot unravel!

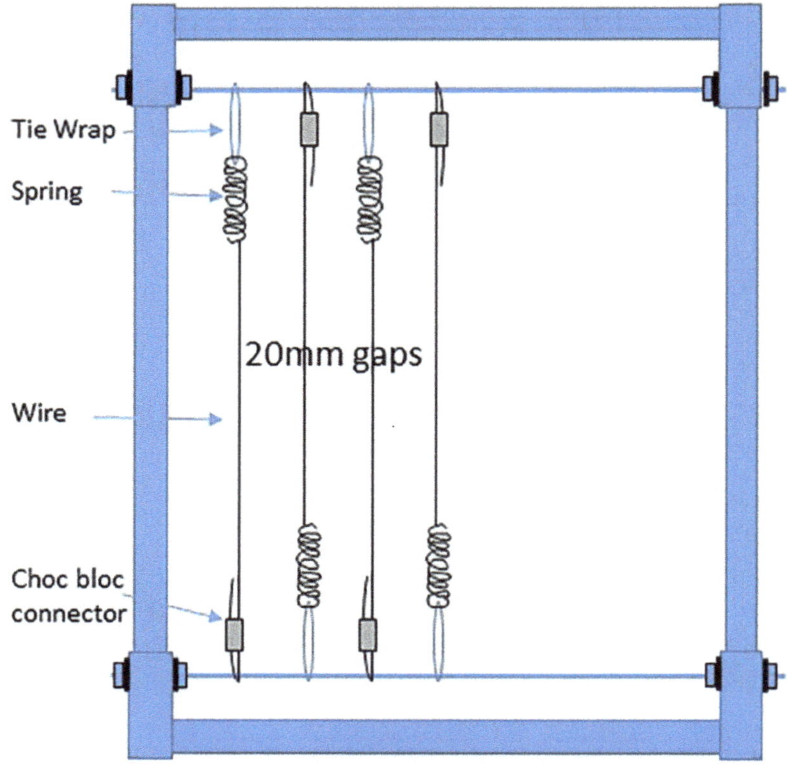

The harp's grid wires assuming you want to adjust spacing at both ends.

- Mark the rods at 20mm centres along length (you want a 20mm gap between the wires). Note: INRAE went for a 24mm gap.

- Insert tie-wrap through spring's hook (see photos).

- Take the tie-wrap over the rod and form a loop.

- Engage the tie-wrap but only just — you will tighten the loop properly when you finally tension the spring at the very end.

- Cut an 1100mm length of s/s wire and tie one end to other end of spring using a non-slip fisherman's knot (see photos). That length will give you enough spare to tie the knot to the spring and to loop round the rod or tie it to the rod, as you choose.

- Take the wire down through connector, round the rod and back up through connector, make sure wire is lightly tensioned (make sure choc-bloc has gripped both wires). Do not cut the wire.

- ▸ Do next wire in that polarity skipping a centre mark

- ▸ Turn the frame upside down and do the other polarity's wires.

- ▸ Now connect each wire in that polarity with the next one's connector using the tail that you have left.

- ▸ Take the tail and bend it gracefully out round the adjacent wire so it is clear of it (under the tie wrap's tail). When it is the right length to go into the choc-block of the next wire in that same polarity, bend it back on itself (so you have a double thickness).

- ▸ Slacken top screw (only) of choc-bloc and insert the doubled-over tail into it, then retighten screw.

- ▸ Make sure the screw has gripped the wire.

- ▸ In addition to any electrical connection made by the wire going round the rod, you have now a set of tight connections, linking all the wires in that polarity together.

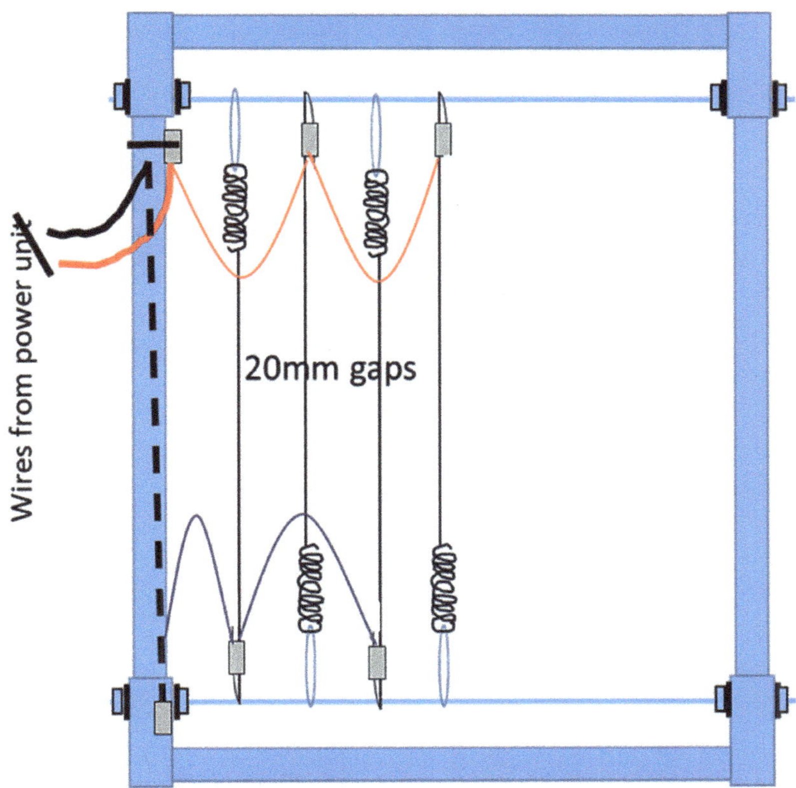

Connecting jump wires

Some Reminders and Photographs:

Your biggest enemy working with steel wire is to get a kink in the wire because you will never get it out. At all times keep the wire under control on the reel and do not allow it to unravel itself; keep it within the reel. If it unspools do not try to pull it straight – it will kink!

When tying the wire to the spring use a knot such as that used by fishermen to attach nylon line to a hook (see sequence of photographs below).

Take the tail through the hook (1), under the wire, then back over (2) and feed the tail under and through the loop (3). Use needle-nose pliers to pull the tail tight whilst holding the wire and get a tight knot on the spring (4).

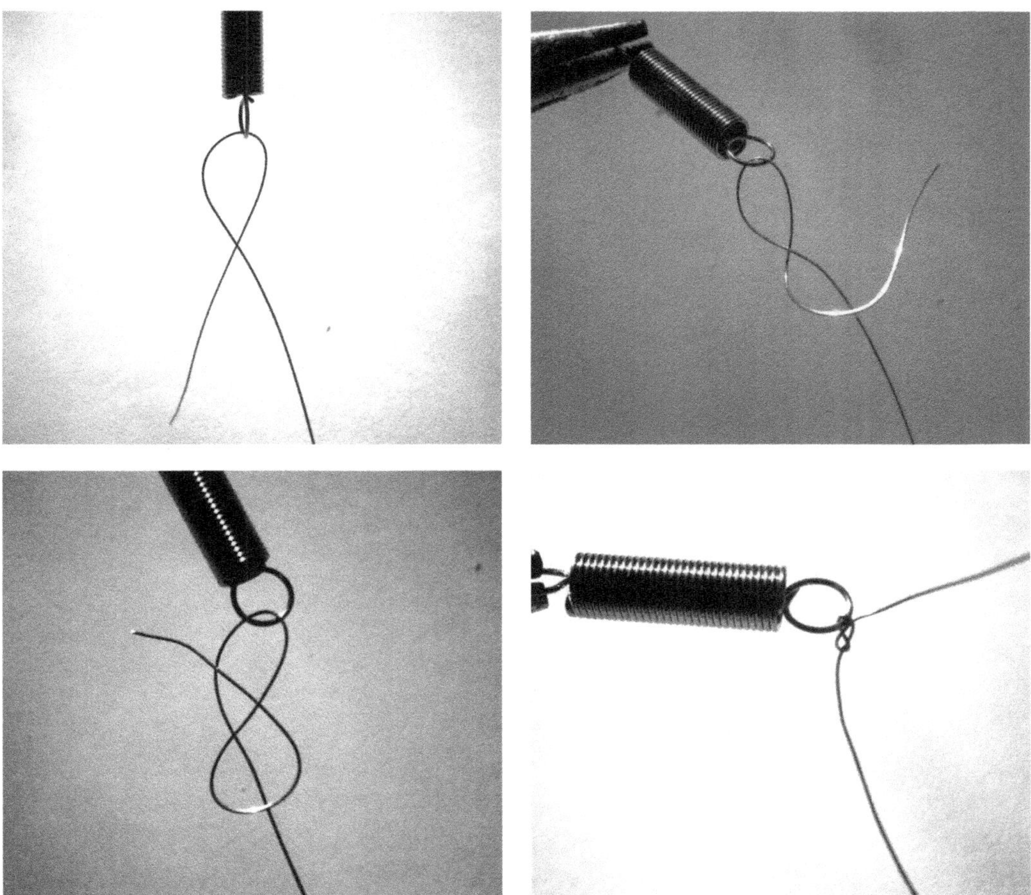

Taking a tie-wrap around the bar (below). Put the tie-wrap through the other end of the spring and just engage it by a couple of clicks. Do not tighten it at this stage. Do not trim the tie-wrap tail.

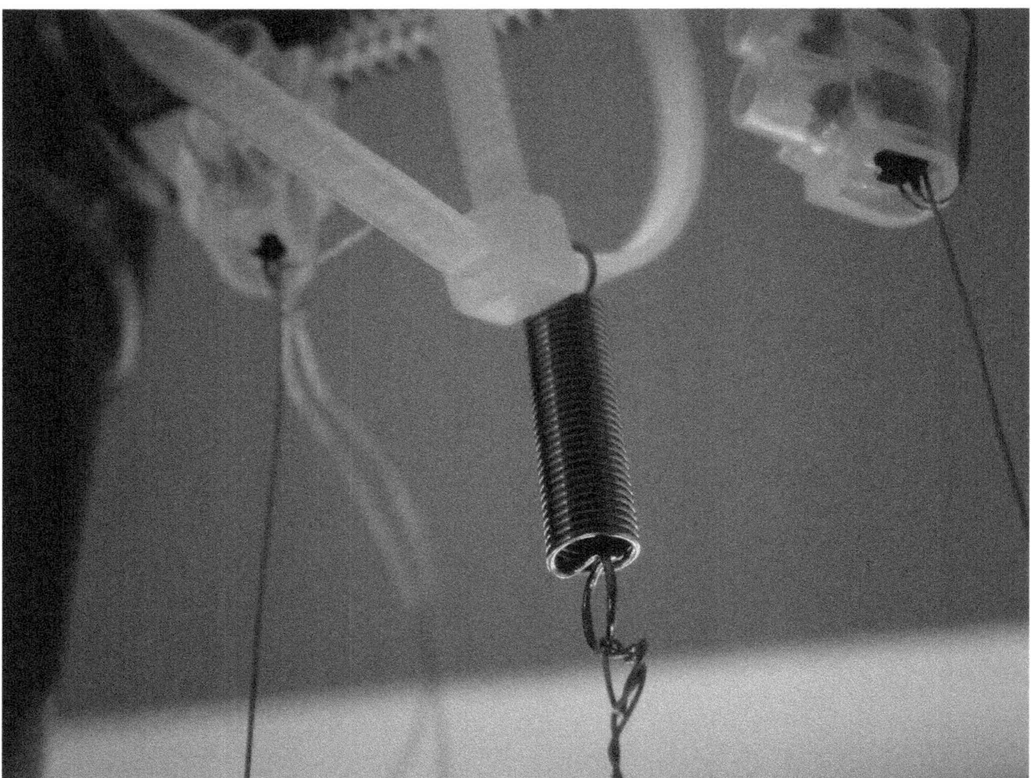

Now to the other bar (in this case bottom bar). Feed the wire through the choc-bloc, under the bar then back up through the choc-bloc. Making sure that you do not tighten the choc-bloc down onto the bar (you need a loop round the bar that is big enough to move the loop along the threads for final adjustments), pull the free end until the spring is slightly expanded and without loosening it so the spring closes up, screw the choc-bloc screws tight.

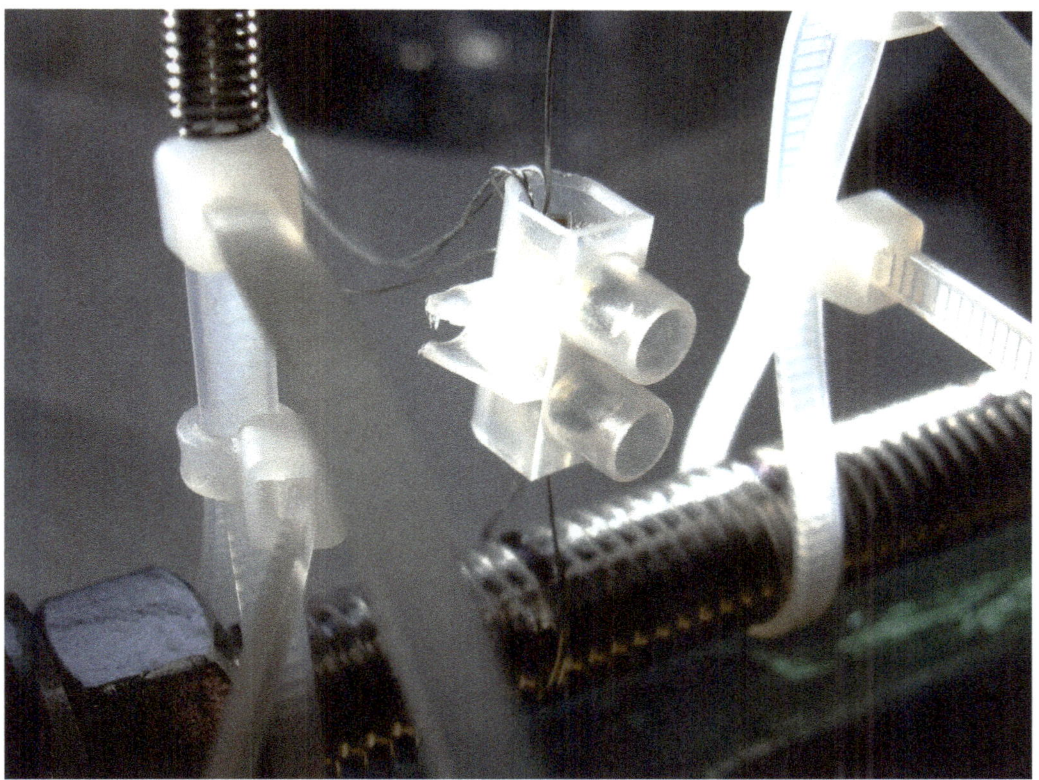

Do not trim the tail because when all the wires are fixed you are going to take the tail across to the next wire in that polarity and connect that into the choc-bloc of that wire.

Harpe electrique – 12Volt Timer and Voltage Conversion Unit (12V to 5V USB outlets)

Overview

This unit connects to a 12-volt car battery and supplies low voltage power to the harp's HV unit, which in this case has a 5V USB feed.

The programmable timer ensures that the harp(s) is/are only powered when required (i.e., daytime).

The voltage converter steps the voltage from a 12-volt supply down to the 5 volts required. Incorporated into the circuit is an in-line fuse holder fitted with a 5 Amp fuse.

Components required

- 1 x Pair of crocodile clip battery connectors
- 1x In-line 12-volt blade fuse holder with 5-amp fuse
- 1 x 12-volt DC programmable timer (source Amazon) Budget £13
- Mini CN101A Time Switch Large-size LCD Digital Microcomputer Control Power Timer Switch High Precision DC 12V Anti-interference
- 1x 12 volt to 5-volt USB voltage converter with two USB female outlets (Amazon) Budget £7.99.

  (RUNCCI-YUN Car Power Converter DC 12V to 5V/ 3A Voltage Converter with Dual USB Adapter Connectors)
- USB to Micro USB fast charge 4.5 metre leads (Amazon) Budget £8 ea.

  Micro USB Cable, 4.5M Extra Long Android Fast Charging Cable, High Speed Micro Sync USB Cable Compatible for PS4/PS4 Pro / PS4 Slim Controllers, Samsung Galaxy S6/S7, Xbox, Nexus, Sony Phone and More.
- Box & cable gland and various 12v crimp connectors (blade type for timer)

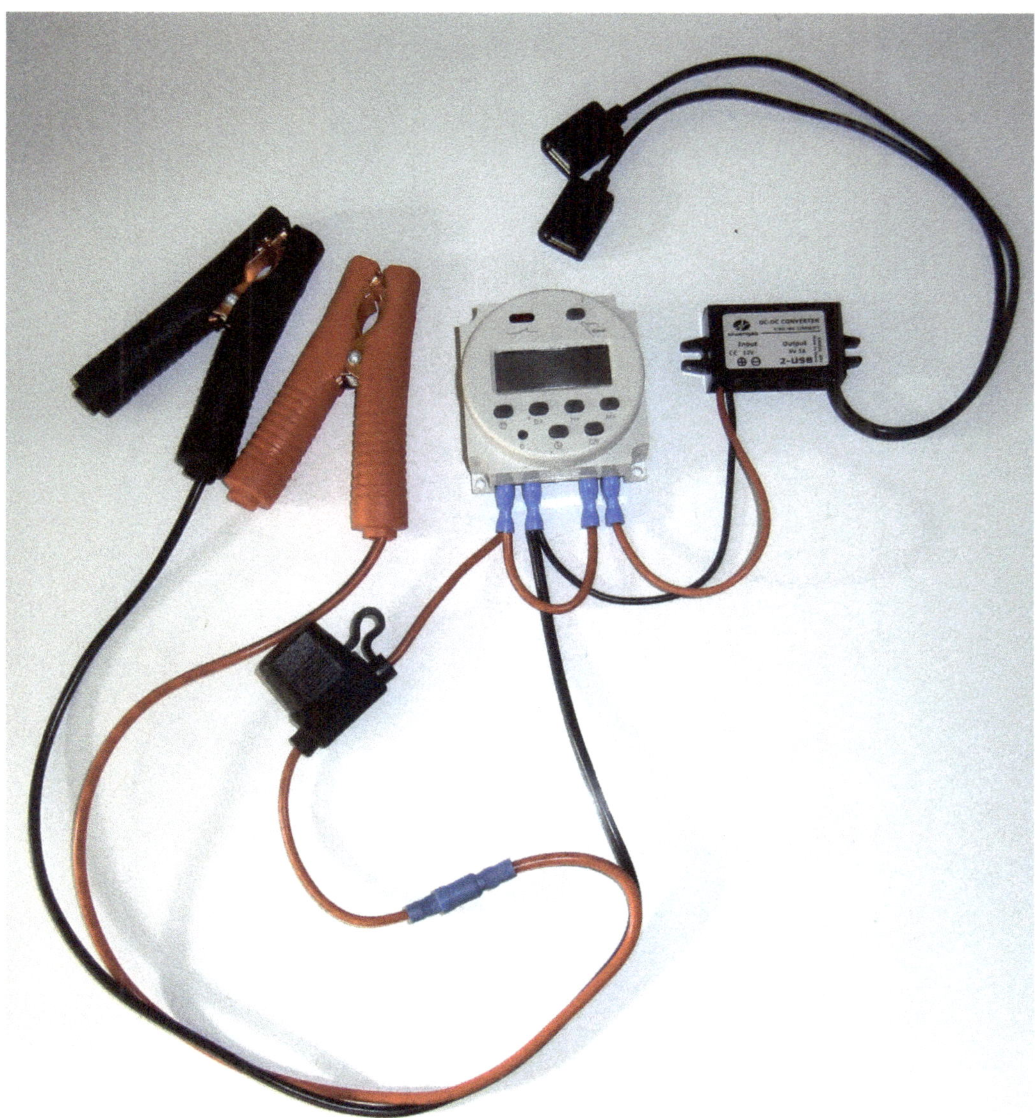

Wiring

Always have a fuse in the circuit to the battery. Put the wires through the cable gland and into the box first (the photograph above is just to illustrate the wiring). Put USB cables through first and then the battery wires.

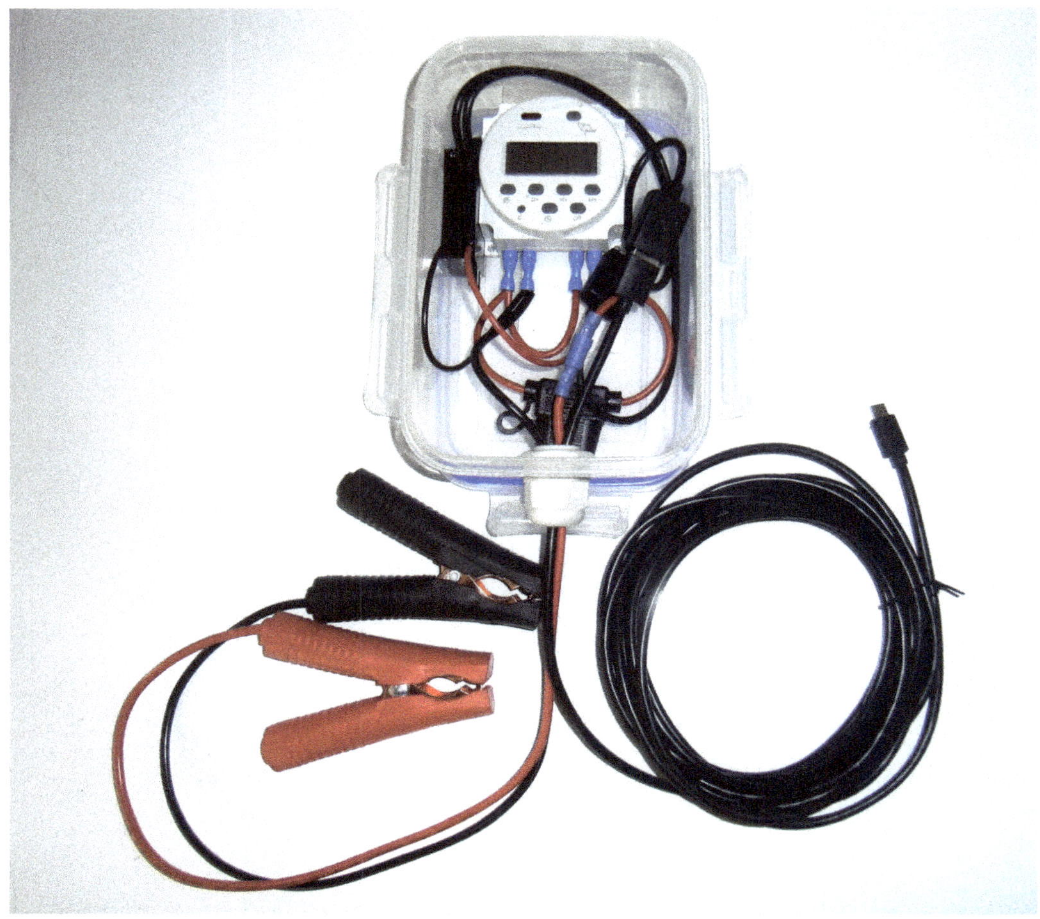

N.B. The box is adequate for 12V but <u>NOT</u> for 240V (mains voltage)

Harpe electrique – High Voltage Power Unit

Overview

The HV power unit provides high voltage to the harp's grid of wires. The unit is powered via a 5-volt USB cable that is plugged into a separate 12-volt to 5-volt converter, controlled by a 12-volt timer, connected via a fuse to a 12-volt battery.

The power unit is made by adapting an electric fly swatter (Zapper bat).

N.B. It has to be the version of the Zapper Bat (or an equivalent) that is rechargeable using a USB lead. Do not buy the much cheaper 'battery only' version, that would be no good at all.

Warning

Even at a low current, the high voltage generated by the power unit presents a risk of electric shock and unlike an electric fence, the output is constant i.e., not pulsed, so it is potentially more dangerous.

You are about to dis-assemble a manufactured item in order to misuse it for a purpose for which it was not designed and in the process you will remove safety features. **Do not proceed unless you are competent to do so and understand the danger.**

If you proceed, you do so entirely at your own risk!

Components required

▶ Qty 1 x ZAP IT! Electric Fly Swatter - Rechargeable Fly Zapper, Electric Fly Killer Racket, Mosquito And Wasp Killer Bug Zapper - USB Charging, Lightweight Handy Fly Zapper (LARGE, 1-Pack) – Amazon £22

▶ Electrical tape

▶ Two lengths of 8-amp wire in red and in black

▶ Qty 1 x 5 amp 'choc bloc' connector (3-way)

▶ Box to accommodate 25cm bat (Sistema spaghetti container - Amazon £7) & cable- gland

▶ You will need a small cross head screwdriver, wire cutter and stripper, small flat-head electrical screwdriver.

Step One – Ensure the Zapper bat is off and use an electrical screwdriver to short-out the bat's grid in order to ensure the unit's capacitor is fully discharged.

WARNING – Always check the unit is turned off and is fully discharged before handling by shorting out the wires on the grid using an insulated screwdriver.

Step Two – Take the Zapper bat and remove the 4 cross-head screws – three in the rear of the bat and one under the yellow cover to the USB charging socket.

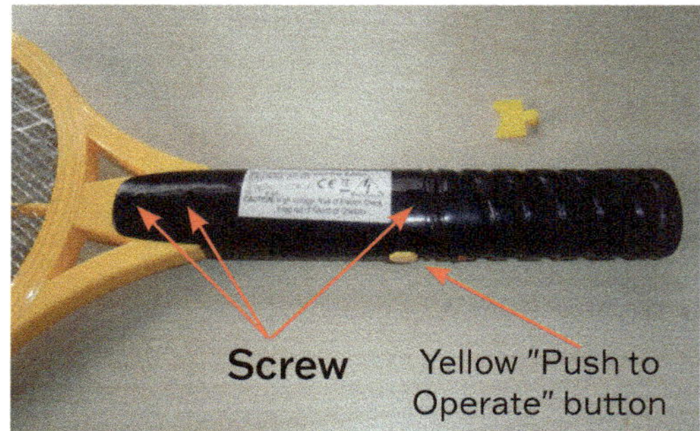

Step Three – Take the Zapper bat apart. Make sure you do not lose the small yellow cover over the 'push to operate' button. Unscrew the two screws holding the printed circuit board (PCB) in place (arrowed below). Carefully prise the board off its spigot and unplug and remove the battery. Do not leave the battery connected; there is a risk of over-charging the battery if you leave it connected and the only source of power should be via the USB lead. Carefully replace the PCB making sure the LED's are correctly located in their holes.

Cut the blue wires and the red wire to the bat's grid (if you prise the bat's head apart you can gain more wire to work with).

WARNING – The battery must be disposed of in accordance with local regulations. Do not dispose of it in the household waste. NEVER SHORT OUT A BATTERY and do not expose to heat or water. Risk of fire and/or explosion!

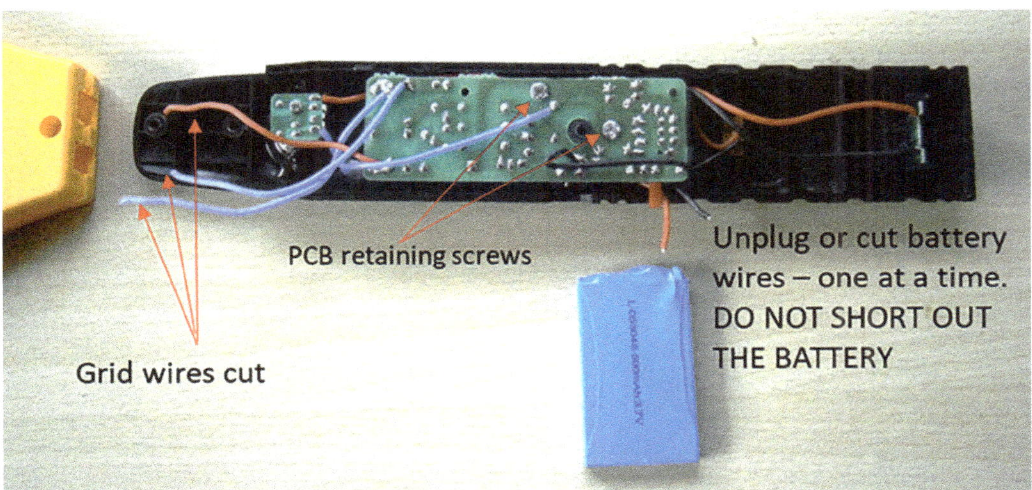

Step Four – Connect the two blue and one red wire into a 5-amp 'choc-bloc' connector as shown. These will be connected to the grid of the harp. Ensure the wires that led to the battery are taped off if you didn't unplug it.

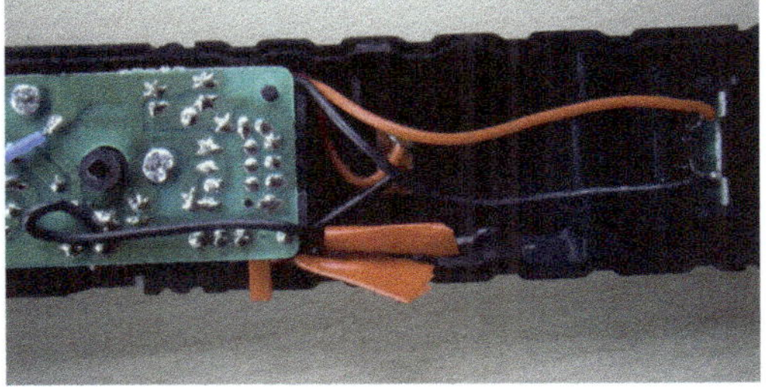

Step Five – Make sure the yellow cover to the 'push-to-operate button' is correctly located and that the USB socket has not been displaced. Make sure that you are not going to trap any wires and replace the cover and screws.

Step Six – Tape the wires to the handle and tape down the 'push-to-operate' button. Place in a suitable box (e.g., a 'Sistema' spaghetti box) ready to connect to the harp.

WARNING. You have just removed one of the manufacturer's safety-features because now both of your hands could come into contact with the high voltage output.

Note: In the photograph above the HV feed wire to the grid is a two-core wire. Although it worked fine for me, two single feed wires would reduce the risk of voltage leakage between the two wires.

The last tail on the harp's grid will be unconnected so now you can add a connector to that and you are ready to connect the power unit's wires. Consider routing the negative wire inside the tube?

Lastly, use a 20mm spacer to check the gaps between wires and adjust the tie-wraps and/or wire loop around the rods.

The supply voltage to the insect zapper bat may be only 5 volts but an output of 3000 volts will give you a nasty shock and could kill you if the current crossed your heart.

Electric *harpes*/*arpa*'s and their HV supply units are potentially dangerous. Do not touch any of the exposed wires and do not move a unit that is switched on. Even when disconnected at the battery or switched off at the On/Off switch, the capacitor in the HV unit can retain its charge for a short period so it important to discharge the unit before handling by shorting out two of the grid wires with insulated pliers.

My recommendation is to incorporate a dual-pole weatherproof switch into the feed to the grid so that you can make sure the grid is off before handling the harp.

ENSURE THE APPROPRIATE WARNING SIGN IS PLACED ON THE UNIT & HARP

Keep the HV supply leads to the harp's grid separated or they will leak current from the positive wire to the negative and you'll get very little output at the grid. Ordinary cables might be designed to insulate up to 400V but they won't stop leakage at 3,000V. It is quite possible with the more powerful French HV power units to get arcing between supply wires if you are not careful!

Powering the *Harpe(s) Electrique* in the Apiary

The high-voltage units that power the grid-wires need a power supply, unless you have bought a very expensive unit that can take a 230V supply from the mains. Most units take either a 12-volt supply or a USB supply.

Most apiaries do not have an outside electrical supply and so the HV unit will have to be powered by a battery (typically a 12-volt leisure battery) that may be supplemented by a solar panel. Be careful about buying a solar powered unit designed for Spain or southern France where the sunshine is stronger and more consistent. I suggest a panel with a 25-watt output for the UK but even then, the battery may need to be periodically recharged off-site especially after a week of overcast skies.

Calculate the power demand of the equipment you wish to power in the apiary. You need to know the total load in amperes (A) or milliamperes (mA). A typical harp HV power unit could be anything between 100mA to 150mA, to which you must add the timer's power load (say 100mA). Add up the total load and check that any converter(s) can power that load.

Now you need to know what that load is going to be in terms of amp/hours over a period of a week (see last page ' Photovoltaic panels and batteries – doing the maths…').

It may be that a standard 12-volt car battery of around 40 AmpHr's will power everything for a week or more, especially if a 12V electronic timer is used (Amazon). The beekeeper may then choose whether to alternate a second battery or recharge the depleted battery overnight. This could be a much more cost-effective alternative to a full solar power set-up (See Inset below – 'Photovoltaic panels and batteries — doing the maths')

If the decision is made to use solar power to supply several harps, the set-up would look something like the diagram below and we will look at each component in turn (the HV power unit/generator was considered above).

WARNING

Solar Panels deliver DC voltage, which is more dangerous than AC. For DC voltages, currents above 25 mA at 50 V are considered hazardous under normal conditions

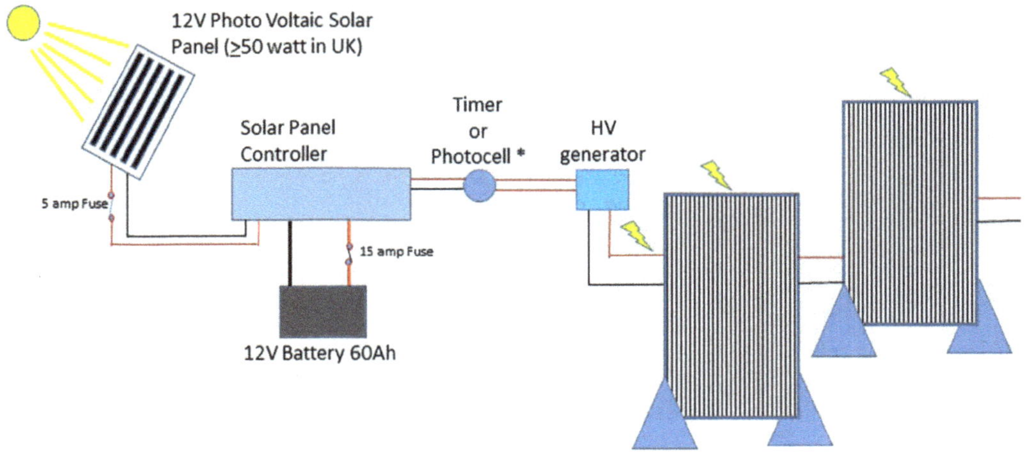

- Power supplies must always be fused.
- Care must be taken to follow component manufacturers' instructions.
- High voltages are potentially dangerous.
- Photo-voltaic panels generate DC (direct current) – take extra care to observe wiring and earthing instructions.
- * Photocell unit must turn "off" at dusk and "on" at dawn (may be included with the HV generator)

Diagrammatic Typical Set-up

The Photovoltaic Solar Panel

Photovoltaic panels are readily obtainable and we would be looking for one rated in watts at a 'nominal' 12 volts. 'Nominal' because the voltage delivered by the panel varies widely dependent on the strength of the sunshine. The voltage variation should not be a problem if the panel is connected to a 12V volt battery, the voltage of which is also nominal depending on the charge state of the battery. But if the HV equipment is connected directly to the panel and especially if the panels are providing too much power, then the equipment can be damaged. In this instance it is wise to use a solar panel controller (see below).

Two ECO-WORTHY 25-Watt 12 Volt Polycrystalline Solar Panel Modules providing solar power – Photograph by Andrew Durham (Note: One panel should suffice; a second can easily be added if required)

The Solar Panel Controller

There is an extra bit of kit that is usually required when operating solar panels, especially of 25-watt rating or more, and that is a solar charge controller. This controls the supply from the solar panel to the battery and protects the battery from over-charging and over-discharge.
Take specialist advice if you decide not to use one.

 A Sunix® 20A 12V/24V Solar Charge Controller (£16.99 from Amazon) (note the addition of a 12V timer (right) and two 12v-5V USB converters (left) from Amazon. There are 4 x 5V USB outlets (far right) and a 12V output, all feeding harps. The USB sockets on the controller itself are not used because they are not controlled by the timer– Photograph - Andrew Durham

The importance of a timer...

There is no point in powering a *harpe / arpa* from dusk to dawn because the Asian hornet doesn't fly after dusk, so you may as well save your battery. Furthermore, there is the risk that nocturnal flyers such as moths and bats could fly into the wires. A 12V timer is readily available from Amazon. Alternatively, a photocell may be incorporated but make sure that the photocell does switch off at dusk and on at dawn, most stand-alone photocells are designed to do the opposite!

Solar Power

Photovoltaic panels are rated using a standard testing condition (STC) and the power is stated as watts.

However, the rated output can only be achieved if:

- The sun is shining!
- The panel is directly facing the sun.
- The sunshine is not diluted by haze or thin cloud
- If the panel is set to face south, which is the norm, the sun will be shining on it at an oblique angle for much of the day and this can reduce the panel's output.
- The panel should be mounted at angle of between 30-45 deg.
- N.B. Any shade on any cells in the panel will reduce output.
- It is best to plan on less than 50% of the rated output.
- Sunshine hours are recorded for locations around the UK and for example, Norwich has some 200 hours in August, 125 in September and 100 in October; the months in which the *arpa electrica* would be deployed.

  It is wise to always use the lower figure in such calculations i.e., 100 hours or an average of 3.3 hours of sunshine a day.

Photovoltaic panels and batteries – doing the maths...

The specification for your *harpe / arpa*'s high voltage power unit will state:

- The supply voltage e.g., 5V (USB), 12V, 24V etc.
- The current drawn from the power supply, stated in amperes (A or amps) or you will have to measure the power draw with an ammeter.
- The discharge voltage at the grid (V) – we can ignore that from now on.

The battery will state its nominal voltage (V) and the capacity of the battery to supply that voltage over a number of hours (Ah).

Photovoltaic panels state their output in volts (V) and electrical power is stated in watts (W).

Example calculation:

The high voltage power unit(s) require(s) a supply of 12volts and will draw from its supply a current of 0.5 amperes total (e.g., 4 harps plus timer).

The battery must be a 12V battery (common in cars). If the battery is rated at 40Ah at a 1Ahr draw (amp-hours) it can in theory supply the HV power unit for 80 hours (40Ah/0.5A= 80 hours) but see below.

If the battery's draw rate is not cited then a reasonable assumption is that the manufacturer is using the standard 20hr-rate i.e., the amps that will discharge the battery is over a period of 20hrs (known as C20). E.G., 40Ah battery at C20 will support a 2 amp draw for 20hours.

However, one should not normally use more than 50% of the battery's capacity or risk ruining the battery, so its useable capacity is 20Ahr (or 40 hrs supply to the HV power unit circuit at 0.5 amps).

The photovoltaic panel is rated at 25W (watts) at 12V (volts). In ideal sunshine conditions this will deliver a power of 2 amps (25W/12V= 2.08A) but this is its maximum output and we would be wise to rely on half that figure – say 1 amp, although that is being conservative (see inset panel Solar Power).

Our apiary's location is in Norwich which is recorded as having an average 125 hours of sunshine a month in September but only 100 hours in October (the two months when we really need our *arpa* to operate). We must use the October figure. We have ensured our system is on a timer so that it only operates for 10 hours a day in September and 8 hours a day in October.

The calculation will be:

HV power unit will demand 0.5A x 8hr/day x 31 (days) = 124Ahr total draw in October. The photo-voltaic panel will provide an assumed 1A x 100hr = 100Ahr total output in October (leaving a nominal deficit of 24Ahr).
A battery will be required for backup. We must also allow for a week when the sun is not shining brightly enough to provide sufficient output from the panel alone.

Practical Solution: We have a safety margin on the panel's output and could go to a 75% battery discharge if absolutely necessary. So, use a 25W panel and 40>Ahr battery, one can always add a second solar panel or increase the battery capacity, if required.